Arms and Innovation

Arms and Innovation

Entrepreneurship and Alliances in the
Twenty-First-Century Defense Industry

JAMES HASIK

The University of Chicago Press
Chicago and London

James Hasik is a consultant to the aerospace and arms industries. He is the coauthor of *The Precision Revolution* and has been an expert commentator on international security topics for CNN, the *Boston Globe,* CBS Radio, and Australian National Radio, among others.

The University of Chicago Press, Chicago 60637
The University of Chicago Press, Ltd., London
© 2008 by The University of Chicago
All rights reserved. Published 2008
Printed in the United States of America

17 16 15 14 13 12 11 10 09 08   1 2 3 4 5

ISBN-13: 978-0-226-31886-8 (cloth)
ISBN-13: 978-0-226-31886-9 (paper)

An earlier version of chapter 2 and some text in chapter 1 first appeared as James Hasik, "Dream Teams and Brilliant Eyes: The SBIRS Low Program, Northrop Grumman's Acquisition of TRW, and the Implications for the Structure of the Military Space Industry," *Defense & Security Analysis* (March 2004): 55–67. Reprinted by permission of Taylor & Francis, LTD, http://www.tandf.co.uk/journals.

Library of Congress Cataloging-in-Publication Data

Hasik, James M., 1966–
    Arms and innovation : entrepreneurship and alliances in the twenty-first-century defense industry / James Hasik.
        p.   cm.
    Includes bibliographical references and index.
    ISBN-13: 978-0-226-31886-8 (cloth : alk. paper)
    ISBN-10: 0-226-31886-9 (cloth : alk. paper)   1. Defense industries—United States.   2. Defense industries—Technological innovations—United States.   I. Title.
    HD9743.U62H37   2008
    338.4'735500973—dc22

                                                        2008013574

This book is dedicated to
Lieutenant Colonel Byron E. Beal, USAF
1951–1998
who fought against the Ba'athists in the 1991 campaign
with his F-4G Wild Weasel and the High-speed
Anti-radiation Missile (HARM)

Contents

The Fast and the Many
The Theoretical Background on Small Firms and Alliances in the Arms Industry

[Based] on current trends, the weapons of the 2030s might be roughly as bright as chickens. Chickens may not sound too fearsome. But these will be single-minded, fully networked, hunter-killer cyberchickens—with strength in numbers. Thanks to civilian demand for smaller, quicker, better computers, they will also be relatively cheap.

GREG CANAVAN, *Los Alamos National Laboratory*[1]

The armaments industry is widely regarded as technologically intensive and often innovative. Since the end of the Cold War, it has also been the subject of several waves of consolidation, in most cases encouraged by the national governments that are its primary customers. Today, the industry appears dominated by a handful of large firms with annual sales of ten of billions of dollars: Boeing, the European Aeronautic Defense and Space Company (EADS), Lockheed Martin, BAE Systems, Northrop Grumman, Raytheon, General Dynamics, Finmeccanica, and Thales. Indeed, there are good reasons for their success. The effects of scope and scale have increased with technological advances in weapons design and production. The larger firms have the resources for large marketing and lobbying campaigns; they have huge research and development staffs; and through the constant migration of military officers from active to retired service, they are generally assumed to have strong grasps of their customers' needs.

Small Business in the New Defense Industry

It may be thought remarkable, then, that some of the most innovative weapon systems that have debuted in the past ten years were developed by rather small businesses, small at least by the standards of the industry. This book is a study—a series of case studies—of six such combinations: small organizations that won contracts to develop or produce important, innovative systems for the U.S. Department of Defense and other military customers. The question of whether small or large firms should be better at innovation is an old one. Over seventy years ago, the economist Joseph Schumpeter distinguished between two agents and coordinating forms of innovation: small firms and entrepreneurial innovation, and large firms and managed innovation. Schumpeter, backed by the later writings of John Kenneth Galbraith,[2] argued that large firms were necessary for sustained innovation in the midst of clusters of technological advancement. The relative innovativeness of small versus large firms has been observed to vary considerably with economic sector and even over time. At one point, small pharmaceutical firms were hardly known as major sources of advances. More recently, the development of pure research organizations considerably changed that perception. The same has happened in the computer field. Where most innovations were once in the capital-intensive business of hardware research, a considerable share of today's innovations is in software, where small firms sometimes excel with little more than "two programmers and a coffee pot."[3] Individual industries also show migration back and forth from the entrepreneurial innovation of small firms to the managed innovation of large firms, and back again.[4] This dichotomy, however, is no longer effective, as hybrid forms of innovation organization, involving universities, pure research organizations, venture capitalists, entrepreneurs, and conglomerates, interact in complex innovation networks, exchanging ideas, labor, and capital in novel ways.[5] Alliances of small firms in the armaments industry have been hatched in large part to combine entrepreneurial drive with economic wherewithal, but managerial coordination was sometimes lacking with this approach.

Innovation, of course, is not merely a matter for laboratories and proving grounds. Changing the very structure of an enterprise may spur innovative activity if it brings together the right staff under the same roof. In this context, mergers and corporate acquisitions are being frequently used in the arms industry today as a means to acquire external[6] or complementary technologies.[7] In the mid-1990s in the United States, a time and place of intense mergers and acquisitions (M&A) activity, acquiring firms in high-technology industries tended to have lower research and development (R&D) intensity than acquired firms, which suggests that the acquiring firms were out to buy technologies that were not available through market transactions or internal development, or to preclude competitors from obtaining these technologies.[8] This should not be seen as technologically backward but rather as innovatively efficient: if centralized R&D is more cost-effective, then less of it may be required for the same effect. In a second-order effect, this newly increased technological capability can help firms recognize new, external technologies that may present opportunities for them. Technology planning in armaments must therefore be an essential element of corporate planning.[9] In many industries, the people who plan corporate strategy differ quite significantly in skills and orientation from those who work directly in technological development.[10] Fortunately, this may be less so in the aerospace, military electronics, and arms industries given their rather technological nature. Strategic planning is important, if not essential, for the growth and development of even small firms in industries experiencing consistent innovation.[11] It is the larger firms, however, that more often have these managerial and analytical capabilities.

Further, the question of who pays for innovation in the short run can have a great effect on who is successful at innovation in the long run. Today widespread disagreement exists over the drivers of military-technical innovation—whether these are endogenous features of the military requirements and funding processes, or exogenous artifacts of purely technical innovation.[12] If the drive for innovation is endogenous to military demand, then the military itself is rather responsible for determining who will be innovative by deciding who will get the R&D funds. If innovation is driven primarily by industrial supply, then firms with a nose for exogenous technical developments have an advantage. This would seem to favor those with larger involvement in commercial products or at least those who employ a significant number of engineers with commercial experience or sensibilities. This may include computer systems firms with significant commercial operations, such as Computer Sciences Corporation (CSC), but it is less likely to include firms that are almost entirely focused on military and security markets, such as Lockheed Martin.

More specifically, despite the aforementioned "allegations of Galbraith, Schumpeter and others, there is little evidence that industrial giants are needed in all or even most industries to insure rapid technological change and rapid utilization of new techniques."[13] Indeed, some extremely concentrated industries are extremely uninnovative, such as tobacco, steel, liquor, shipbuilding, newspapers, plate glass, and meatpacking.[14] Steel and shipbuilding have shown improvements in innovativeness of late, but lagged for decades. In computer software, a variety of models exist, and some sectors are extremely concentrated, while some of the most profound innovations of recent years have been launched by small teams of researchers working at the fringes of large organizations, and sometimes, outside them entirely.[15] Indeed, there is reason to believe that the introduction of computer software as an essential factor of production may be changing the dynamics of innovation and their effects on industrial structure. Before the development of computer-aided software engineering (CASE) tools, virtually all mechanization concerned manufacturing processes—not design or management.[16] Since these are now on the table as well, it is possible that, to use the stock phrase of Internet enthusiasts, everything has changed.

Product development has economies and diseconomies all its own. For example, Lockheed Martin and its customer, the U.S. Air Force (USAF), suffered cost overruns and labor constraints on the F-22 Raptor stealth fighter due to a decision taken in the 1990s to build the highly integrated suite of onboard computers around the Intel 960 processor. Intel was actively marketing the chip's design to the banking sector because its multiple layers of security built into the hardware would alleviate many of the banks' computer security concerns. The 960, however, failed to catch on with commercial customers, so Lockheed Martin was left with an orphaned technology that required extensive training for every new computer programmer joining the Raptor project.[17] Had the project been smaller or perceived as less important to the future of the USAF, it might have been scuppered for such miscues. Since it was such a valued endeavor and technological stretch, the cost overruns continued.

This brings us directly to the questions of firm size and dynamic efficiency, and the theoretical groundwork for our case studies. Partly because small and medium-sized enterprises are held in some endearment by a great many people in Europe and North America, a large body of literature has arisen to describe the industrial conditions under which small firms are likely to have advantages. These include industries with three characteristics. The first concerns innovation, but in several ways: *small firms are often relatively advantaged in industries that are highly innovative but that are low in R&D intensity, and in which uncertainty about markets or future technological trajectories is high.*[18]

The first condition, innovation, clearly applies to much of the arms industry. Innovation is a difficult concept to characterize rigorously, but not all technological advancements, however profound, are equally innovative. The MQ-1 Predator combat drone (described in chap. 3) is generally not as capable an attack aircraft as the F-18E/F Super Hornet fighter-bomber, but it would be difficult to argue that the latter aircraft is more innovative, even if it is, in some aspects, more technologically advanced. This is not to say that only innovative systems are militarily valuable or that the most marginally valuable military systems are always innovative. Innovation must be defined as distinct from other forms of incremental development in order to identify the technologies and markets in which smaller enterprises are likely to thrive.

Low R&D intensity, is prevalent in some, but not all, segments of the arms industry. Since the experience of the Second World War, armaments R&D has captured popular imagination as essential to victory.[19] Money poured into military scientific research during the war, and the results were seen in radars, electronic navigation, jet engines, and the atomic bomb. Sine the late 1950s, firms producing guided missiles, combat aircraft, and military electronics have been the recipients of large and sustained government investments in technical research. This does not mean that all segments have had an equal degree of R&D investment. Military R&D spending was sharply reduced in North America and Europe following the end of the Cold War, and some sectors suffered more than others. Spending on stealth aircraft research remained strong but dropped precipitously for new armored vehicles. R&D intensity also changed within sectors. The U.S. Army spent $251 million on basic research in helicopter technology in 1984, but only $115 million in 2004. The National Aeronautics and Space Administration (NASA) shut down its two helicopter wind tunnels and its only helicopter crash test facility as well.[20] Eurocopter, however, invested rather heavily in its rotor-quieting technology, which enabled the company to capture the largest share of the commercial and civil government helicopter market.[21]

Uncertainty arises because the innovation in question is radical (vice incremental) and because the dominant products in these situations are often relatively early in their life cycles. Uncertainty also arguably characterizes many sectors of the arms industry since the end of the Cold War. The dissipation of the well-defined threat of the Warsaw Pact meant that many national governments lacked clarity about the exact purpose of many segments of their armed forces, and the amorphous nature of the transnational threats that arose further complicated force structure planning. Hence, the arms markets have been subject to considerable questions not just about technological trajectories but also about whether technologically possible weapons would be politically suitable. As the ensuing case studies will show, the decade of the 1990s witnessed considerable innovation, in part because arms procurement budgets had declined so sharply, and in part because of the widespread availability of one infrastructural technology in which the Pentagon had invested so heavily—Global Positioning System (GPS).

Why do small firms tend to be relatively advantaged under these conditions? While large firms have the Schumpeterian benefits of scale and scope, small firms are often more nimble organizations which can more quickly make decisions on which technologies and systems to pursue. If the innovation required does not depend on high science and large laboratories, the process of discovery and development can all the more easily be undertaken without large capital outlays. If the particular product trajectories are uncertain, small firms are often more willing to take the risks with customers and their own product lines that are necessary to bring new products to market. Large, diversified defense contractors may actually suppress sensible weapon and platform concepts for fear of cannibalizing sales elsewhere or annoying factions of officers and officials who oppose the warfighting concepts that they represent.

The second condition favoring small firms concerns skill: *Small firms often have an advantage in industries that require a high proportion of skilled labor, and in which the production is more skill-intensive* than capital-intensive. This rather characterizes almost every corner of the arms business, but it characterizes some to a greater degree than others. Munitions production is relatively capital-intensive and can be accomplished by production workers who are not necessarily degreed engineers. Space satellites, on the other hand, are essentially built by hand, one at a time. Shipbuilding has a similar dynamic, though it is more capital-intensive in its production, has more scope for somewhat less skilled labor, and features somewhat discontinuous scale economies. In speaking of scale and scope economies, that is, one must clearly distinguish among both scale and scope economies and continuities at the product, plant, and corporate level.[22] Not all will aid the largest of firms, and these discontinuities can provide economic cover under which smaller firms can thrive.

Small firms should not be thought more likely than large ones to have skilled staff, but they often use them in a more focused fashion. Modular product architectures allow the decomposition of production into relatively atomized supply chains in which each assembly or subsystem is made by a specialized organization. In many cases, those are small firms, since the scale of production required is relatively small, but also because small organizations can more easily be tight-knit ones in which management can more readily attain the particular mix of skills needed to most effectively develop and field innovative products. The importance of skill thus points out the nature of those sectors in which smaller enterprises may be relatively advantaged. If the systems of systems under consideration today are built with relatively open physical and informational architectures, and relatively heterogeneous collections of platforms, weapons, and sensors, then smaller firms may have an important role both in competition and in conjunction with larger ones. As individual subsystems can be conceived, defined, and developed in a relatively separable manner, highly innovative and skill-intensive firms may be found more capable of rapidly developing new technologies that require only limited production runs.

A third condition involves learning: *Small firms have an advantage in industries that are composed of a relatively high proportion of large firms, but that also have room for smaller competitors. Industries often feature this relatively heterogeneous structure when their products are subject to medium-speed learning curves—that is, the learning-by-doing of production is neither particularly rapid nor slow.*[23] This matter expands on the preceding point regarding skill. Even if the arms industry seems to be dominated by giant firms there are many sectors in which small producers command considerable customer loyalty in seemingly secure niches. Virtually all of the military helicopter market may be divided between the five aforementioned firms, but MD Helicopters has often been mentioned as a notable competitor. Formerly a division of Mc-Donnell Douglas, and recently purchased by the investment firm Patriarch Partners, the company is small by industry standards, with a single plant in Mesa, Arizona. Since the Vietnam War, however, it has been making various versions of the MD500 Defender, which is rather beloved by the U.S. Army's commando helicopter force as the MH-6 Little Bird.[24]

Production niches often feature a particular set of industrial dynamics. If learning-by-doing in the sector is quite rapid, then large firms may have an advantage because the methods of production are likely to be (as noted above) relatively capital-intensive, and larger firms often have better access to more cost-effective capital. If learning is too slow, then little rapid progress is possible, and acumen in marketing, distribution, and large-scale production will be more important. The computer processor industry exemplifies the former situation and soap powder the latter. While subject to radically different design and production learning curves, both are relatively highly concentrated among a few large firms. Niche markets are very profitably served by small firms, but effective entry on a moderate scale is unusual. Many sectors of the arms industry lie in this middle space. Production quantities are often too low for high-speed learning curves to take effect: while automotive manufacturers build cars by the tens of thousands per year, the entire production run for most armored vehicle models stops at a few hundred or thousand. Production runs are even shorter in combat aircraft.

Cyclicality in the industry is important as well. One might theorize that production methods should be updated during downturns in business because the relatively idle staffs will have more time for installation and training. It is at least as common that firms invest in capacity when upswings in demand justify the outlays. Another model argues that in many cyclical industries, firms update their production methods during upswings in business because production staff can more effectively learn by doing when they are actually doing the work of production at a faster pace.[25] During sharp downturns—from which immediate recovery is not expected—the departures of considerable portions of the skilled staff can bring about organizational forgetting as well. Former Defense secretary William Perry's "Last Supper" message in 1993 made clear that the peace dividend of the 1990s was expected to be long lasting, so retrenchment and restructuring were the primary organizational concerns of the larger arms makers for much of the 1990s. In some of the cases I will discuss, this provided an opportunity for smaller manufacturers and software houses to forge ahead with new and unexpected technologies.

Overall, the advantage of the small firm in these situations is said to de-
rive substantially from its organizational flexibility—a smaller organization is
simply not as bureaucratic as its larger competitors and can thus respond to
changes in technological or market conditions more rapidly. The technological or organizational paradigms into which researchers at large organizations
sometimes fall can be rather "blinkering"—that is, they can "have a powerful
exclusion effect: the efforts and technological imagination of engineers and the
organization in which they work are focused in rather precise directions, while
they are, so to speak, 'blind' with respect to other technological possibilities."[26]

In many cases, it is the breadth of these technological possibilities today
that provides the opportunities. The rapid, long-range, and networked operations that define modern military capability generally do not arise from
focusing on the production of individual aircraft and weapons. Rather, this
sort of capability requires attention primarily to the large-scale integration
of multiple systems and capabilities. Thus, the requirements of information
age warfare mean that today's reasons for restructuring may be quite technological.[27] Mastery of the sort of military information described above is the
province of only a few countries around the globe, and it dominates military
considerations much as nuclear weapons did during the Cold War.[28] However, as political scientist Andrew Latham has written, the innovation that
produces these systems "is increasingly based on the dynamic recombination
of generic technologies," which are often information technologies.[29] Single
firms focused on particular niches rarely find all the technology and know-
how they need to produce these systems within the walls of the corporation.[30]
This means that systems integration, not production, is often the most important competency for a defense contractor seeking to provide the best and
widest service to its customers. By the late 1990s, roughly 75 percent of the
U.S. Department of Defense's procurement budget was already being spent
on aircraft, missiles, electronics, and communications equipment.[31] Take out
the airframes, and the Pentagon might be viewed as today spending the bulk
of its equipment funds on precision strike networks. Stitching those networks
together would logically be the next challenge.

However, the tendency toward increased scope on the part of the largest
armaments firms has opened an opportunity for smaller, innovative firms.
If technological innovation is increasingly based on dynamic recombination,
then the ability of arms producers to predict the course of technological
innovation will further diminish in the future.[32] Interconnections between
the technologies will be important, but the basic technologies of precision
engagement and networked intelligence, surveillance, and reconnaissance
(ISR) will be critical factors in fielding new capabilities. If these systems are

increasingly modular, then smaller firms are likely to gain a relative advantage in producing them. In specific (postparadigmatic) industries, suppliers are at least as often the sources of innovation as primary producers. Industry leaders are vulnerable to substitute products.[33] Even if the armaments industry cannot be described as postparadigmatic, the new dynamics of innovation point to important technologies originating in smaller enterprises. Over the long run, therefore, industry leaders could become vulnerable to the predations of smaller firms, and unsophisticated vertical integration may very well become less effective.

The Force-Multiplying Role of Alliances

The ensuing series of case studies will identify the circumstances under which small firms in the arms industry will be relatively advantaged. The discussion will also illustrate how and when firms should consider alliances as a means to leverage their best attributes while minimizing their limitations. This is important not just for small firms but large ones as well. Large firms have coordination costs, and these only grow with size. Some industry observers have particularly faulted Raytheon and Lockheed Martin for their inability to properly integrate into their operations the many companies that each purchased in the merger waves of the 1990s. Alliances (or their equity form, joint ventures) have gained currency over the past few decades as an intermediate form of production organization between pure markets and the purely hierarchical organization of integrated companies.[34]

However, many of the oft-cited reasons for alliances among independent companies do not withstand economic scrutiny in the long run. *The sharing of risks and costs* are often cited as reasons for creating alliances. An aerospace project can often require more capital than even a large company's equity base—even Boeing brings in partners on new jetliner projects.[35] However, an integrated firm would avoid the coordination costs of an alliance, and would thus, *ceteris paribus*, be more efficient. This is thus relevant to the European "national champions" of the 1970s—which continuously sought alliances with one another to take on large armaments integration projects—but much less so to larger U.S. and pan-European firms today. Risks and costs are in any case considerably less in military development programs than in commercial ones, since the taxpayers foot most of these bills.

Economies of scale, as suggested above, arguably drove much of the alliance activity among European aerospace firms over the past forty years. Since public procurement in North America and Europe has often shown a soft spot for small and medium-sized enterprises, smaller firms have often been encouraged to form alliances with other small firms, or even large firms, to compete for projects that would otherwise be too large for them to tackle alone. They have been a particularly popular form in the European aerospace industry, albeit for interstate political reasons that are not fully replicated on the other side of the Atlantic. However, with the considerable consolidation of the industry over the past fifteen years, this motivation should be ebbing.

Access to markets is also a common reason for alliances, particularly cross-border ones in Europe. In a transatlantic example, the alliance between General Electric (GE) and Snecma Moteur in jet engines—the joint venture company CFM International (CFM)—was initially largely about Snecma's relationship (frankly, one of subsidiarity) with the French government, and GE's relationship with U.S. aircraft manufacturers Boeing and McDonnell Douglas.[36] The move was arguably successful. While rival Pratt & Whitney held 90 percent of the market for new, large jet engines in the 1970s, by 2000, CFM held 51 percent to Pratt's 11 percent.[37] This is not to say, however, that market access can be always bought so cheaply. Projects such as the Space-Based Infrared System (SBIRS) involve technology so sensitive that the national governments will often hesitate to share it with any other government, except for possibly in the case of the Anglo-Saxon ABCA grouping (the United States, Great Britain, Canada, Australia, and by extension, New Zealand). The initial resistance that Lockheed Martin and AgustaWestland encountered while offering their EH101 Merlin helicopter as a Presidential transport in the United States is a testament to how tight markets can be, often without good reason.

Conversely, despite the enthusiasm expressed by many small firm coalitions, the costs associated with organizing production on an interfirm basis are clearly understood. These include:

Contracting costs. These include the cost of negotiating, monitoring, and renegotiating an agreement should conditions change in a significant and unforeseen way.[38] As we shall see in the case of the alliances for SBIRS Low, these costs were likely to be considerable, unless many of the fifteen members of the TRW/Raytheon team were treated more as subcontractors than as allies. This may have been the case, but a TRW/Northrop Grumman combination would remove these costs between two of the more involved alliance partners.

Knowledge spillover. The more partners in an alliance, the greater the chance of a spillover of knowledge. This has a particularly deleterious effect on coordination costs in a long project.[39] Obsession with unauthorized skills transfers hampers the effectiveness of teams.[40] With several different software and hardware contractors on each side, the individual firms in large alliances often cannot be assured that their tacit knowledge or even proprietary methods will not leak to firms that could be competitors.

With costs and risks as these, one can say that an alliance of companies will only be *ex ante* more efficient than a single, integrated company if it offers the participants a superior combination of risk and return through a "peculiar combination of scale and flexibility advantages."[41] This will most likely be the case when the two firms in question are operating under several conditions.[42] The first is *considerable change with respect to processes and goals.* If goals and processes were stable, then integration into a single company would pose fewer risks, and the full reduction in transaction costs through integration would accrue to the combination. Unstable processes and goals, however, pose significant risks for any company expanding beyond its base of skills. This has definitely characterized many military procurement programs, particularly in the technical frontiers of unmanned aircraft and orbital weapon systems but even in such theoretically stable fields as medium-weight armored vehicles.[43]

A second condition for the effectiveness of an alliance of companies is *moderate appropriability of intellectual assets.* Here we are referring again to the aforementioned leakiness of knowledge. Too much appropriability indicates a market solution, or no solution at all—firms have little incentive to engage others that will probably steal their technology. Indeed, research suggests that many companies actually underinvest in research and development because they are uncertain about how to prevent the benefits from leaking out to the entire industrial base. While beneficial for the customer (the government) in the short run, this is bad in the long run, as it discourages innovation by contractors.[44] Weak appropriability suggests that an integrated company would be the better approach: cooperation between firms would be too difficult, since there would be little way to transfer knowledge, which is largely the point of cooperation in the first place. In the weapons business, appropriability is often rather low. So much of the information about design and production is so classified that even sharing information willingly within a firm is difficult. In theory, this alone would, *ceteris paribus*, render alliances rather rare in the business. On the other hand, one might argue that because the government lays claim to much of the technology that is developed, it can direct contractors to share data at its will. This, however, suggests extraordinarily close program management, which, as I will show, was not quite the case with the SBIRS Low program. It was also not the case in the Joint Direct Attack Munition (JDAM) program, but the result was happier as the single contractor, Boeing, was very well organized for success.

A third condition facilitating collaboration is *moderate expropriability of quasi-rents from specific assets.* Expropriability refers to the degree to which another party can take control of the proceeds of an economic activity.

Quasi-rents are the near-term earnings above long-term equilibrium profits that can be expected from relatively scarce man-made capital. That is, until
potential competitors enter a market, the incumbents can earn higher, though
not quite monopoly, profits.[45] Specific assets are those that are useful for one
sort of economic activity but generally not for another. Thus, in layman's
parlance, we are referring to a shakedown. The owner of an asset can get
taken by his allies if the value of his assets is considerably lower elsewhere
than in his alliance, and particularly if he also faces significant switching costs
or exit barriers.[46] This has some applicability in the arms business—in many
cases, relationships in partially closed national markets are very stable but
also imbalanced among contractors.

Properly executed, an alliance or joint venture should not be merely a
hybrid governance structure—indeed, one with higher costs than integrated
firms[47]—but an endogenous factor of production in its own right.[48] That is,
alliances should not be entered into because they are fashionable but because
they offer better risk/return ratios to shareholders, which will only happen
if they offer better service in the long run to customers. Collaboration is
indeed particularly pervasive in rapidly changing and knowledge-intensive
businesses,[49] especially in the pre-paradigmatic phase of technological development as customers sort out what sort of products they seek. Collaboration
in an alliance becomes less efficient as products become standardized—the
market risks of corporate bureaucracy abate somewhat, while the advantages
of production efficiencies become more acute.[50] Fortunately for smaller firms
hoping to tap into the value of alliances, many military systems are neither mature nor commoditized products, and military markets will remain
knowledge-intensive and subject to considerable technological change. This,
however, only indicates that alliances are useful in the industry, not that all
alliances are useful. Whether among small firms or between firms small and
large, alliances are not without disadvantages. After I present the six case
studies that are the subject of this book, I will return to the question of when
alliances are appropriate and when small firms have an initial advantage.

The theory regarding small firms has been widely discussed in the general economic literature and is drawn from a number of sources. The theory
regarding alliances was developed by Alexander Gerybadze, professor of international management and innovation at the University of Hohenheim in
Stuttgart. My aim here is to show how these principles apply to the arms
industry and how the changes afoot in military requirements are, contrary to
the received wisdom, relatively increasing the importance of small firms and
alliances. The six case studies are of well-known military systems for which a

small business conspicuously served as the prime contractor or for which a large business notably did when the opposite might have been expected.

Chapter 3. One of the firms selected as a potential prime contractor by the USAF to develop its SBIRS Low missile tracking satellites was Spectrum Astro, a relative newcomer to the industry with a staff of under four hundred. Along with its competitor, the relatively large TRW, it failed to meet the USAF's expansive technical requirements in a timely fashion, so the program was restructured into a sole-source award to a team of contractors chosen by the program office.

Chapter 4. The Predator reconnaissance drones that proved so useful over Bosnia, Kosovo, Afghanistan, Iraq, and elsewhere were initially designed by a team of roughly a dozen engineers at a small, virtually unknown division of a nuclear reactor controls firm in San Diego, General Atomics.

Chapter 5. JDAMs systematically blasted the Taliban from Afghanistan in 2001 and were dropped by the thousands on Iraqi forces in 2003. The JDAM is the first affordable, autonomous, adverse-weather precision weapon ever developed. While it was conceived and prototyped by a rather small team of researchers at the USAF Weapons Laboratory in the early 1990s, it is in serial production by Boeing.

Chapter 6. The Royal Australian Navy rather stunned the world with its speed in deploying heavy equipment to East Timor for peace enforcement in 1999. Its catamaran transport, HMAS *Jervis Bay*, moved twice as fast as conventional monohull ships and cared little about the state of port facilities. The ship was built by a car ferry maker in Tasmania whose name, Incat, was little known in naval circles.

Chapter 7. The PowerScene mission rehearsal and planning system (MP&RS) proved very useful in Bosnia—both in bombing Serbian installations and afterward in facilitating the Dayton Accords negotiations. It was developed by a small software firm in Virginia, Cambridge Research Associates, which was later knocked out of the business by Lockheed Martin.

Chapter 8. Cougar and Buffalo mine-resistant vehicles were lifesavers for explosive ordnance disposal teams in Afghanistan and Iraq from 2004 onward. As of this writing (March 2007), though roadside bombs were the leading killer of U.S. troops in the two campaigns combined, it appears that only one Coalition soldier had been killed in either of those two vehicles. Both were developed with technology licensed from the South African government and built by Force Protection Industries, a South Carolinian company that had fewer than ten employees when the Afghan war started in 2001.

These cases are drawn from six of the most important segments of the arms business: space satellites, combat aircraft, guided missiles, shipbuilding, computer software, and ground combat vehicles. All represent relatively innovative weapons concepts pursued by relatively small product development and production staffs. Three—the Predator, the Cougar, and the military catamarans like the *Jervis Bay*—represent successful products for the small or medium-sized enterprises—General Atomics, Force Protection, and Incat—that produce them. One—PowerScene—was a very successful product in the short run, but its success did not translate into success for the company, Cambridge Research, that produced it. Another—the JDAM—was a successful and very innovative product, whose concept was hatched by a small team, but it became a money-maker for Boeing, the world's largest aerospace company. The very first case to be discussed—SBIRS Low—was a technologically ambitious program undertaken by a rather small firm for its industry, Spectrum Astro. The program foundered and was repeatedly restructured, and Spectrum Astro was eventually acquired by General Dynamics, one of the world's largest arms makers.

This also raises the question of what constitutes a small firm. To many small business owners, an enterprise with a staff of a thousand is not a small company, but this the criterion that the U.S. Small Business Administration (SBA) uses to sort companies in many economic sectors according to the North American Industrial Classification (NAIC) scheme.[51] The table below indicates the size of what the SBA considers a small firm by the number of staff in the sectors in question in our the six case studies, and the size of the actual firms in question. All six firms treated in the case studies meet this general criterion, with one particular exception: Boeing and JDAM are included to illustrate when a small business might not have a reasonable shot at winning a contract.

SBA SMALL BUSINESS SIZE STANDARDS BY NAIC SECTOR AND THE SIZES
OF THE FIRMS IN THE SIX CORRESPONDING CASE STUDIES

Industrial Sector	Size Considered Small	Case Study Company	Size of Company
Spacecraft	1,000	Spectrum Astro	400
Aircraft	1,500	General Atomics	1,000
Guided missiles	1,000	Boeing	156,000
Shipbuilding	1,000	Incat	300
Information technology value-added reselling	150	Cambridge Research Associates	40
Industrial truck manufacturing	750	Force Protection	400

Whatever the measure, small arms firms have an important role to play in the cycles of military-industrial development, but the conditions under which they thrive must be thoroughly understood. Certain aspects of the ongoing military-technical revolution and the changing nature of military requirements have opened the door for smaller firms to innovate in areas in which larger firms may not so readily do. Small, innovative firms are often the subject of great interest—by larger firms seeking to acquire their technologies, skills, and entrepreneurial cultures; by entrepreneurs seeking to emulate their success; and by government procurement and industrial policy authorities seeking to nurture their success. These cases bear lessons for all three groups.

There is one further matter to explain: sample size. This study of the industry considers individual cases of weapon systems and the firms that manufacture them. This has been done for both expediency of analysis and richness of detail. It is true that a highly quantitative study of hundreds, if not thousands, of firms would elicit vigorous nodding by economists for the rigor of its approach and the presumed robustness of its results. "Rolling the tape" with event studies of stock prices is appealing, but this approach is rather difficult in armaments for three reasons":

- Government customers are not always forthcoming about the prices paid for large, particularly classified systems. More so, some governments actively hide some of their purchases to keep their capabilities a mystery or to hide affiliations with producers.
- A greater problem lies with the nature of the data available. Corporate accounts can provide some information, despite the malleability of accounting standards, but the highly diversified operations of some the largest arms producers will obscure information about individual sectors of activity. Worse, government accounting systems are generally designed not to produce managerially useful information at all, but instead to prevent civil servants from stealing money.[52] Thus, in an industry whose activities should be very transparent—in which public funds procure the majority of the equipment—price and cost data are actually quite opaque.
- Lastly, many governments spread purchases between firms or at least favor firms whose design and production facilities are domiciled on their territory, for a variety of reasons both related to national security and internal distributive politics. Whether this is viewed as industrial policy, supplier management, or political patronage, it does demonstrate the analytical problems of monopsony: since governments are often the only legitimate buyers of heavy weaponry, the market does not function as a classical market, and thus requires more subtle tools of analysis.

If, though, arms firms are indeed extremely heterogeneous and market segments are many and overlapping, then the case study method can be very enlightening. Case studies peel away layers of effects to uncover patterns of truths between programs, products, firms, and industrial sectors. If statistical studies are more reliable at telling us what is happening, then case studies are often more effective at telling how and why it is happening.[53] For managers, marketers, and engineers in the arms industry, this is arguably at least as important.

Dream Teams and Brilliant Eyes
The SBIRS Low Program, Northrop Grumman's Acquisition of TRW, and the Implications for the Structure of the Military Space Industry

We've sort of become the leading underdog of the satellite guys. We're the company shaking things up. Frankly, this industry has a history of poor performance and cost overruns. The Pentagon wants to change that.

DAVID THOMPSON, *founder and CEO, Spectrum Astro*[1]

Introduction

In the summer of 2002, Northrop Grumman, the third-largest arms manufacturer in the United States and the fourth largest in the world, acquired TRW, a car parts maker more valued for its expertise in lasers and space satellites. Northrop Grumman's proposed acquisition of TRW was intended to produce a single company with stronger skills in space satellites, missile defenses, and electronic intelligence systems. The integration of the two companies—billed as "relatively simple" by management—would produce a firm better able to compete on a wide range of large projects against Lockheed Martin, Boeing, Raytheon, British Aerospace, and EADS.[2] Those large projects were those that would, by the company's estimation, produce the "systems of systems" that were supposed to transform the Pentagon's military machine into an unassailable twenty-first-century force.[3] Large systems, the argument goes, require large contractors, so a combination with TRW would be in the best interests of the Department of Defense, if only to increase the ranks of the largest U.S.-based contractors from two to three.

In this context, an excellent case study for the *ex ante* value of the proposed merger is the U.S. Air Force's (USAF) Space-Based Infrared System Low (SBIRS Low). A development of the earlier "Brilliant Eyes" program, SBIRS (pronounced "sibbers") Low was a plan for a large constellation of ballistic missile tracking satellites. This was probably the only program in which Northrop Grumman and TRW were working so closely that it posed vertical integration issues with antitrust implications.[4] Just before the acquisition was proposed, TRW was a potential prime contractor competing against the rather small firm Spectrum Astro in the project's ongoing program definition and risk reduction (PD&RR) phase. By acquiring TRW, Northrop Grumman was to be, without a significant restructuring of the program (see below), a subcontractor of one variety or another to whichever team won the contract. The complexity of the project and its unusual management structure in the year preceding the acquisition suggested that a Northrop Grumman/TRW combination would improve the program's chances of success.

Spectrum Astro, TRW's competitor, was a firm of less than five hundred people. It was understandably proud—and perhaps even defensive—about its role as a potential prime contractor for one of the USAF's most important satellite programs. Scott Yeakel, Spectrum Astro's Vice President for SBIRS Low, had this to say after the team passed its system design review in April 2001:

> Our team has once again demonstrated that bringing together industry leaders, regardless of corporate size, with each member demonstrating complementary skills and technical expertise, is the future of the defense business and provides our nation with the most innovative space systems at the best value . . . Clearly we showed that we are one seamless team, working without barriers to develop this vital system for our future defenses against missile attacks.[5]

Regardless of the performance of the Spectrum Astro team on SBIRS Low, this seamlessness is not universally present. Were all teams of small companies seamless, there would be far fewer large companies in the world. Indeed, Northrop Grumman maintained that its proposed acquisition of TRW would benefit the government by providing the smaller contractor—widely acknowledged as a leader in space systems and lasers—access to wider technologies of systems integration. TRW might plausibly have asked, however, why the two companies needed to join forces so formally, and why the existing teaming arrangements would not continue to work. After all, the search for complementary skills, especially in technology-intensive industries, is one of the most-cited reasons for companies to enter into alliances with one another.[6]

The Defense Satellite Program (DSP), a SBIRS Forerunner, Would Not Support Missile Defense Requirements

To understand the importance that the USAF placed on the SBIRS Low program and the role of SBIRS Low as a case study of evolving industry structure, some technical history is in order.[7] SBIRS was conceived in the aftermath of the 1991 Gulf War. During the war, a USAF Defense Satellite Program (DSP) satellite in geosynchronous orbit over the Middle East was used to detect the exhaust plumes of ballistic missiles (Scud missiles and modified

Scuds) launched out of Iraq toward Israel, Saudi Arabia, and Bahrain. Though the original mission was to detect longer-ranged launches out of the Soviet Union—launches with larger infrared signatures of longer duration—the DSP satellite detected almost all of the Scud launches in 1991. There were three limitations with the process, however. The first was its manual communications aspect: DSP data had to be passed to commanders in the region by voice through North American Aerospace Defense Command (NORAD) headquarters in Colorado Springs, so the information was not useful for tracking missiles in flight. At best, Patriot missile battery crews could be alerted that a weapon was inbound and the crews could then begin searching the skies with their radars. People on the ground at potential targets could be warned to take cover.

A second DSP limitation involved sensor sensitivity. The DSPs were never designed for tracking shorter-range ballistic missiles. The charged-coupled device (CCD) on the satellites was sufficiently sensitive to detect the three-minute high temperature burn of a Soviet intercontinental ballistic missile (ICBM). Scuds, however, burn for not much more than a minute (depending on range), and with a lower exhaust temperature. The DSP constellation also required relatively fewer satellites and less-sensitive sensors because military intelligence had previously established the locations and missile types of every ICBM field in the Soviet Union. Finally, the DSPs were aided in their task by the Diyarbakir tracking radar in Turkey, which was physically swiveled about to face Iraq—it was normally used for observing launches from the Baikonur cosmodrome in Kazakhstan.[8] Thus, just detecting the Scuds was better performance than required by the DSP satellites' mission.

Third, in addition to its sensors being insensitive, they were also indiscriminate. DSP data could be used for targeting the launchers after the missiles were gone. Since the Iraqis had only ten usable launchers (9P117 MAZs of Soviet manufacture), but fired 93 missiles, destroying a launcher could slow down operations. The problem was that the DSP satellites were never intended for this purpose either. Each carries a single CCD—the rotation of the satellite enables it to scan continuously the entire disk of the earth. The time between sweeps is thus such that the DSPs can only pinpoint a launch to an area about eight kilometers on a side. At night—when most of the Scuds were launched—the USAF sent F-15E fighter-bombers with LANTIRN (Low Altitude Navigation and Targeting Infrared for Night) pods combing the desert for signs of activity. Looking down a LANTIRN pod, however, has been likened to viewing the world through a soda straw: the high resolution of pod comes at a high price in its narrow field of view.[9] Since aircraft equipped with LANTIRN pods required a cue of area not more than a quarter mile on a side, it is doubtful that any launchers were destroyed in "the Great Scud Hunt."[10]

SBIRS: An Ambitious Program from the Start

After this experience, the USAF decided to look into two constellations for detecting, tracking, and *targeting* ballistic missile launches of short to intercontinental range: SBIRS High and SBIRS Low. SBIRS High was conceived as a constellation of four geosynchronous satellites and two SBIRS payloads on other (unspecified) elliptically orbiting craft. The elliptical payloads were scheduled at the time to be available in 2004; the geosynchronous satellites were scheduled to be launched starting in 2006, and the entire constellation was to be in place by 2010. SBIRS High satellites would raster-scan the disk of the earth but would also have staring sensors to focus attention on particular areas of concern (which DSP does not). This would be good for cueing fighter-bombers onto launchers but not as useful for midcourse missile defense (see below). SBIRS Low was conceived of as a constellation of about thirty low earth orbit (LEO) satellites. Launchers were initially scheduled to begin in 2006, and the constellation was to be fully operational by 2011. SBIRS Low satellites would have target acquisition and tracking sensors for cueing weapons onto the missiles themselves, as well as communications cross-links between the satellites so that the constellation could automatically share targeting information without a ground-based relay.

The problem with procuring these systems lies in the challenge of missile defense. In the architecture of the Ground-based Midcourse (missile) Defense Segment (GMDS, formerly called the National Missile Defense system [NMD]), destroying a ballistic missile in this fashion is complicated. The following sequence of events (for defending against a missile launched from the western Pacific rim—most probably from North Korea) illustrates the difficulty: The X-Band radar could be cued with SBIRS High alone, but the tracking basket would be much larger, which means that the missile would present a more difficult target.[11] There are other problems with forgoing the SBIRS Low project. Without SBIRS Low, the future Space-Based Laser (SBL) project would be in jeopardy. The SBL constellation was conceived as consisting of roughly twenty-four laser in low earth orbit that could destroy ballistic missiles before they reached apogee. The weapons would also have a theoretical capability against aircraft, cruise missiles, and surface targets. However, the SBL would absolutely require targeting information from SBIRS Low in order to destroy ballistic missiles in the boost phase. In addition, if the SBL could ever be made to work against terrestrial or atmospheric targets, its considerable geopolitical potential would provide even more reason for getting SBIRS Low to work.

SBIRS High Was Not Doing Well

Since SBIRS was expected to do so much more than the DSP system that it was replacing, one could believe that the program was going to be challenging from the start. Indeed, SBIRS High was the first effort by the Department of Defense to get the Space and Missile Systems Center (SMC) in southern California to work with the National Reconnaissance Office in northern Virginia. These institutions on opposite coasts have traditionally represented the "white" (less classified) and "black" (extremely classified) worlds of military space systems management and carry with them significantly different ways of operating. The agencies cooperated, but not as quickly as some in the Pentagon had hoped.[12] SBIRS High subsequently suffered from considerable scope creep— so much so that delays had almost become "a fact of life" in the E-Ring of the Pentagon.[13] While the top-level requirements had been stable since 1996, subsidiary requirements "weren't properly allocated," according to then–Under Secretary of the Air Force Peter Teets.[14] Next, under the increasingly popular concept of Total System Product Responsibility (TSPR), the system program office (SPO) overseeing SBIRS granted prime contractor Lockheed Martin extreme latitude to do as the company saw fit in managing the program. Asking a single agency to consider the management of the entire life cycle of a system was in many ways a sensible approach. In the United States, this process started in the early 1990s with efforts like the Integrated Weapon System Management (IWSM) initiative, a life cycle management approach that precipitated the merger of USAF Systems Command and USAF Logistics Command into USAF Materiel Command (AFMC). That merger, undertaken to facilitate life cycle management, was a challenging undertaking. By asset value, it was one of the largest organizational mergers in history: at the outset AFMC controlled $160 billion in assets.[15]

In the SBIRS Low program, the SPO limited its activities to managing the money flow, defining remaining requirements (in conjunction with Air Force Space Command), and managing system's operational acceptance testing. The combination of scope creep and poor subsequent program management caused costs to burgeon.[16] The Directorate of Requirements for Force Enhancement-Sensors at Air Force Space Command evaluated and monitored the contractor's proposed solutions to ensure that the stated SBIRS needs were met,[17] but it was not clear who was monitoring the solutions for technical competency and budgetary compliance. By the spring of 2002, the SBIRS High program was about $2 billion over budget and three years behind schedule.[18] At roughly the same time, there certainly were other approaches in use.

Beginning in 1995, the National Polar-orbiting Operational Environmental Satellite System (NPOESS) was under joint development by the Department of Defense, the Department of Commerce, and NASA for military, commercial, and scientific weather observation. From 1997 through 1999, the managers of the program at the Department of Commerce's National Oceanographic and Atmospheric Administration funded a set of sensor development projects with multiple contract awards. In December 1999, these were assigned as directed subcontracts to prime contractor TRW, which assumed shared system program responsibility (SSPR) with the program office.[19] On the other hand, by 2006, the estimated cost of the program had risen from $6.5 billion to over $10 billion, so this may also not be an excellent example of program management.[20]

SBIRS Low Was a Particularly Troubled Program

SBIRS Low presented an even more challenging integration problem. According to the General Accounting Office and the Air Force Research Laboratory, SBIRS Low depends on six critical technologies—without any one of them, the system would not function. Five of these were still considered immature technologies: the two infrared focal plane arrays (for target acquisition and tracking), the two cryo-coolers (one for each sensor), and the cross-links among the satellites (which the Department of Defense had never demonstrated on LEO satellites).[21] Missile defense, however, had become such a political priority that the system experienced (according to an independent review impaneled by the Department of Defense) a "rush-to-failure." Eventually, the schedule-driven program was replaced by an event-driven program,[22] and the schedule was stretched—that is, the USAF stopped insisting on rigid adherence to the development schedule, regardless of the maturity of the technologies. Regardless, by the spring of 2002, the first satellites were still not scheduled to be launched until 2006—fully seven years after the start of the program.

Indeed, this plan to delay fielding preceded the selection of the two contracting teams for the PD&RR phase.[23] Additionally, the SPO cancelled orbital and ground demonstration programs by TRW and Boeing, respectively, as they were already expected to exceed their budget by $416 million.[24] This sum alone was thought to be jeopardizing the program.[25] Cancellation of the demonstration programs did not increase confidence in the likelihood of success for overall system. At this stage, while SBIRS Low was being touted as Space Command's top acquisition priority,[26] the Air Force was incorporating the lessons that it had already learned as an example of what not to do with the upcoming Space-Based Radar (SBR) program.[27]

The SBIRS Low Teams Were Large and Loosely Organized

One could argue that the structure of the competing contractor teams had contributed to the program's troubles. After winning its place in the competition, Spectrum Astro promptly took aboard as team members Northrop Grumman, Lockheed Martin, and Boeing—three of the four largest U.S.-based defense contractors. While one might further describe them as subcontractors, they were formally listed as team members. For its part, the TRW-Raytheon team competing in the PD&RR phase was quite large, with formally fifteen members.[28]

RESPONSIBILITIES ON THE TRW-RAYTHEON SBIRS LOW TEAM

Company	Responsibilities
TRW	Prime contractor, program management, mission systems engineering, spacecraft production, space vehicle integration and test, ground segment
Raytheon	Electro-optical payload design and production, mission data processing, mission management, space vehicle integration support
Northrop Grumman	Mission data processing, systems engineering, space segment support
General Dynamics	Constellation and communications network management, production process support
Ball Aerospace	Electro-optical payload thermal management and acquisition sensor production for Raytheon
Honeywell	On-board processor design for payload and spacecraft
Photon	Phenomenology and target modeling
Sparta	Simulation tool design and development, requirements analysis, software development process support
Rockwell Scientific	State-of-the-art infrared focal plane arrays
Agilent	Test planning and test equipment
Ryan	Military utility trades, target signatures, missile defense interface/concept of operations support
SciTec	Phenomenology and target modeling
MRC	Survivability support
CSC	Discrimination and simulation support
FTI	Calibration operations concept and risk handling plan development, sensor calibration process

A review of these companies' roles on the project suggested deep involvement in design and planning—deeper than would be expected even today in activities such as DaimlerChrysler's North American automotive production or Boeing's manufacture (or, integration, in a sense) of jetliners. Significantly, the design and development of the infrared payload was being managed wholly by Northrop Grumman's "mission integrated product team [IPT]." Lockheed Martin was also part of the mission IPT, developing algorithms and other aspects of the ground system; Boeing was part of this IPT as well, developing sensor components and their associated algorithms.[29] Indeed, both teams competing in the PD&RR phase could be described as large and loose alliances. The Emperor Napoleon once famously quipped that if he had to fight, he preferred to fight against a coalition, as it lacked unity of purpose and command. These same tensions exist in commercial alliances and joint ventures.

Alliances Carry Costs

As noted in the introductory chapter, alliances such as these (or their equity form, joint ventures) have gained currency over the past few decades as an intermediate form of production organization between pure markets and the purely hierarchical organization of integrated companies.[30] Here again, we can rule out some of the oft-cited explanations for alliance formation as motivating the deal. Economies of scale, for example, were not the motivator: TRW and Northrop Grumman had, by Northrop's admission, quite dissimilar businesses. Indeed, Northrop Grumman officials were quoted in April 2001 as trumpeting the value of their teaming arrangement with little Spectrum Astro in particular—this, they said, gave them an edge in the competition, but it certainly was not a matter of capacity.[31]

Indeed, despite the enthusiasm that Spectrum Astro's Scott Yeakel expressed for coalitions, the conditions for success through a production alliance were not present in the SBIRS Low program. Contracting costs were considerable, as both teams came with a considerable number of members with competing agendas. Reviewing the criteria set out in chapter 1, we observe a mixture of possibilities and problems:

> *Considerable change with respect to processes and goals.* If goals and processes were stable, then integration into a single company would pose fewer risks and the full reduction in transaction costs through integration would accrue to the combination. Unstable processes and goals, however, pose significant risks for any company expanding beyond its base of skills. This definitely characterized the SBIRS Low program—as it has many recent military satellite programs. The aforementioned shift from a schedule-driven "rush-to-failure" to an event-driven science project suggested considerable instability in the

program's technical goals. The problem here was that while an alliance might have been the best organizational approach to solving this managerial problem, it may still not have been adequate.

Moderate appropriability of intellectual assets (moderately leaky knowledge). Too much appropriability indicates a market solution, or no solution at all. Too little appropriability indicates that an integrated company would be the better approach: cooperation between firms would be too difficult, since there would be little way to transfer knowledge. As noted in chapter 1, appropriability tends to be rather low in the armaments industry: since so much of the information about design and production is classified, even sharing information willingly within a firm is difficult. The government can theoretically direct contractors to share data at its will, but this suggests extraordinarily close program management, which we know not to have been the case with SBIRS. The various contractors also knew that they would soon be competing against each other for programs like the Space-Based Radar (SBR) and the Space-Based Laser (SBL) programs, assuming that the USAF was able to secure funding for those rather ambitious projects. In that event, some aspects of the SBIRS Low design—such as its cryogenically cooled infrared sensors, which would be important for the SBL—would again be highly sought technical assets. This suggests that the firms might not so readily share valuable information about their subsystems in a way that would allow a relatively vertically and horizontally striated alliance to thrive.

Moderate expropriability of quasi-rents from specific assets (moderate potential for a shakedown). The owner of an asset can get taken by his allies if the value of the asset is considerably lower elsewhere than in the alliance.[32] This could be true for certain payloads on electro-optical and electronic intelligence satellites, but only if the payload manufacturer could not integrate the spacecraft designs itself. If the assets were highly specific to the particular type of production in question (*e.g.*, electronic intelligence payloads), this would indicate an integrated solution. If assets were more fungible among different types of production (e.g., automotive electronics), we might expect a more market-oriented solution in which a prime contractor would keep his subcontractors at arm's length.

In summary, it is arguable that surveillance satellites are neither a mature nor commoditized product, and that the market will remain knowledge-intensive and subject to considerable technological change. This however, only indicates, as noted in the preceding, that alliances are useful in the industry, not that all alliances are useful. Indeed, with sprawling product teams—and one led by a smallish prime contractor—the SBIRS project may have represented a case in which the limits of efficient inter-firm cooperation had been reached. Eventually, the Pentagon and the SBIRS SPO decided just that.

The "Dream Team," Restructuring, and a New Name

Born of the political desire for a mid-course missile defense system, the SBIRS Low program was clearly an ambitious technical challenge. By February 2002, public statements from the Department of Defense had begun to indicate that a radical restructuring of the program was under consideration, possibly one in which "you've totally changed the inside . . . the whole thing has been blown open but the contractor base is still there."[33] Since both contracting teams seemed to be stumbling, the SIBRS SPO began thinking that the project was so challenging that there might be at most one team of engineers in the United States that could handle it.

Later that year, the SBIRS SPO—after high-level consultations within the Air Force Department and the Missile Defense Agency—scrapped the on-going PD&RR competition and instead awarded the contract on a sole-source, noncompetitive basis to a "dream team" of the best contractors for every component technology. TRW, soon to be acquired by Northrop Grumman, was awarded prime contracting responsibility. Plucky Spectrum Astro was named to develop the bus and Raytheon to develop the primary infrared sensor. At the same time, the program was fundamentally restructured. Overnight, SBIRS Low became a technology demonstration program. The new goal was to construct and launch two research and development satellites in 2006; in early 2003, the SPO and Northrop Grumman were still trying to define what would happen after that point.[34] Arguably, the entire program reverted to the one part of itself that had been cancelled as costing too much: the on-orbit demonstration of two satellites. Since then, less ambitious missile defense programs, designed to counter shorter-ranged weapons like the Scud, have been garnering at least as much budgetary attention. Just as the United States was invading Iraq, the Pentagon was requesting $1.4 billion for an additional 108 Patriot-3 missiles, which were being expended at a solid rate in Kuwait against incoming rounds. Today, cooperative development of the Medium Extended Air Defense System (MEADS)—the proposed replacement for the Patriot—continues with the German and Italian defense ministries.[35]

Finally, in November 2002, the name of the SBIRS Low was changed to Space Tracking and Surveillance System (STSS), ostensibly to eliminate lingering confusion with the SBIRS High program (which retained its name and its now superfluous designation "high").[36] The name change, however, was also useful in covering the USAF's retreat from its previously unattainable objectives. Selecting the "best of the best" as component makers was not enough to save the program in its initial and overly ambitious form. Still, technological immaturity was only half the problem with SBIRS: the other half was a lack of oversight by the SPO and those contractors supervising the development efforts. Arguably, the DSP satellites had been the result of a tighter and better-integrated team: TRW had built 23 spacecraft, and Aerojet (also now part of Northrop Grumman) had supplied 23 infrared sensors for them. The relationship lasted over 30 years, and the satellites have exceeded their specified design lives by 150 percent through five upgrade programs.[37]

Next Time, Make That a Northrop Grumman + TRW "Dream Team"

While ultimately a failure, the SBIRS Low program still serves as an excellent case study into the requirements of corporate control, contracting economics, and program management of what was one of the most complex military systems then under development in the United States. Specifically, the outcomes of this and related missile defense programs illustrate Northrop Grumman's strategic motivations for acquiring TRW and TRW's for acceding to the acquisition.

TRW was a company of great technical skill, and it could clearly handle systems integration for established satellite programs. For the DSP program, the company (now as part of Northrop Grumman) still provides "end-to-end systems support." It built and integrated the spacecraft, the successor organization still provides day-to-day technical assistance at Schriever and Buckley Air Force bases, analyzes satellite performance, and does anomaly testing and early on-orbit testing from the Orbital Test Station (which is not actually in orbit, of course). The company also wrote the software that processes, displays, and distributes DSP data to the president and the regional commanders.[38] It is fair to say, however, that the design of that software is somewhat manual and outdated—had it been as advanced as that desired for SBIRS, more rapid warnings might have reduced casualties from ballistic missile strikes in the Gulf War. Upgrading this to the next level—satellite crosslinking battle management technology—required wider skills than TRW had available. This led the company to assemble a large team of contractors around its effort, but there are clear problems with teams that are this large and loosely integrated. After all, research indicates that the commercial success of defense consortia increases with the closeness of the commercial relationship[39] and decreases with the number of firms involved.[40]

It is striking that the genesis of TRW resulted in what was considered at the time over-reaching authority for the contractor. In 1958, Thompson Products merged with Ramo-Wooldridge to form Thompson Ramo Wooldridge (TRW). Thompson had hardware contracts with the USAF's ballistic missiles programs, and Ramo-Wooldridge had recently been selected to provide consulting services, including contractor evaluations, to the USAF. Despite legal assurances that a conflict of interest would not be allowed to arise, the USAF chose to establish the Aerospace Corporation to take over TRW's technical services work, and situated the new nonprofit corporation's headquarters one block from its Space and Missile Systems Center at Los Angeles Air Force Station.[41]

Four decades later, however, the situation had changed. What TRW lacked was close integration of its satellite technologies with the systems integration skills of a larger contractor, a relationship at least as close as its thirty-year relationship with the former Aerojet. The SBIRS SPO was fond of describing the SBIRS constellations and mission control station as a "system of systems." Neither TRW nor Northrop Grumman was chosen by the Missile Defense Agency to form one of its "national industry teams" for GMDS. These rights were awarded to Boeing and Lockheed Martin, contractors of considerably greater scope and scale. TRW was indeed participating as a subcontractor, handling systems integration and engineering for Boeing's team and command and control work for Lockheed Martin's team.[42] Assuming that the returns are better for prime contractors, Northrop Grumman and TRW may have better financial possibilities ahead of them as joint participants in bidding for future projects like the SBR and the SBL. Assuming that the government does not wish to repeat its SBIRS Low experience, the Pentagon might appreciate closer management in the future from both its satellite SPOs and single, integrated contractors.

Spectrum Astro and the Question of Size

The remaining question, then, is that of size. Would a company like Spectrum Astro necessarily find itself in over its head with a project like SBIRS Low? Spectrum Astro was certainly the underdog from the start: its selection as a PD&RR contractor in August 1999 over industry heavyweight Lockheed Martin relatively "rocked the space world."[43] Spectrum Astro rather reveled in the distinction as a small and successful firm among large ones with problematic customer relationships: the company defined its beginnings as residing in the promise of "smaller, faster, better, cheaper spacecraft that work." The allure of the small, entrepreneurial, innovative, successful firm is undeniable, but the successful part is the issue. *Business Week* magazine had profiled founder and CEO W. David Thompson, a former USAF space officer, in one of its "entrepreneur profiles" as running "the rising star of aerospace."[44] Reviewing the criteria from chapter 1, we find a mixed bag of indications concerning the potential for this model in the space satellite business.

Innovation without R&D intensity. The first problem is that most of the military space satellite business has been very high in R&D intensity over the past several decades. The entire "cheapsat" or "lightsat" movement within the Department of Defense and NASA arose out of frustration with the idea that every new program had to break great technical ground in a large, expensive, multifunction satellite. Spectrum Astro had made its name and fortune with clever, innovative, and cost-effective designs, not through enormous leaps in basic science. In the mid-1990s, when agencies like NASA and the USAF Research Laboratory demanded efficiency, Spectrum Astro delivered a series of scientific research satellites that met the agency's needs without breaking its budget. Success, however, depends on the continued attraction of clever but not quite path-breaking designs over the pull of the big craft.

Skill-intensive production. Spacecraft are basically hand-built: production lines make little sense when a large program includes only several dozen craft. The USAF's Navstar GPS has been by far the largest program in decades, but the first three flights of primary production (Blocks II, IIA, and IIR) consisted of just forty-nine satellites. The SBIRS program would also be large by satellite standards, but still would only constitute about thirty spacecraft. Building a satellite bus and integrating its components require very smart engineers but little in the way of production facilities besides clean rooms, specialized tooling, and relatively sophisticated computer software.

Medium-speed learning. Spectrum Astro was a very accomplished firm by the time it won a role in the SBIRS Low PD&RR effort. The company had built a well-regarded group of one-off spacecraft for research and testing purposes, particularly as test beds for missile defense and for astronomical research (see the table). However, this meant that Spectrum Astro was learning between programs, because it was learning from programs, but that it did not effectively have learning curves within individual programs. In that respect, the company was attempting a rather great leap forward with SBIRS Low. Until 2004, the company had never sent into orbit a satellite weighing more than a ton. Since weight is so closely watched in spacecraft design, total launch weight is a reasonable if broad analog for the complexity of the spacecraft's design. The DSP satellites would weigh over two tons each.

All the same, alliances between space satellite companies large and small retain popularity. In April 2006, BAE Systems and the storied small firm Surrey Satellite Technology Limited (SSTL) announced an alliance to bring SSTL's small satellite technology to U.S. government customers, particularly NASA and the Department of Defense. Moreover, whatever its merits as an independent operation, Spectrum Astro's core capabilities were not in question. In 2004, General Dynamics (GD) acquired the company and announced that it would form an important part of GD C4 Systems, a division with a staff of

PLANNED AND ACTUAL LAUNCHES OF SPECTRUM ASTRO SPACECRAFT,
1992–2007, AS OF ACQUISITION BY GENERAL DYNAMICS

Satellite	Launch Date	Launch Primary Mission and Date	Total Launch Weight (kg)
GLAST	Feb. 2007	Gamma-ray observation	4527
Swift	Spring 2004	Gamma-ray and x-ray observation	1463
C/NFOS	Jan. 2004	Ionospheric scintillation forecasting	378
Coriolis	Jan. 2003	Microwave radiometric meteorological experimentation	817
RHESSI	Feb. 2002	Gamma-ray and x-ray observation	197
MightySat-II	July 2000	Hyper-spectral imaging	125
Deep Space 1	Oct. 1998	Comet and asteroid reconnaissance	486
MSTI-3	May 1996	Missile detection and tracking experimentation	212
MSTI-2	May 1994	Missile detection and tracking experimentation	169
MSTI-1	Nov. 1992	Missile detection and tracking experimentation	150

Note: RHESSI = Reuven Ramaty High Energy Solar Spectroscopic Imager; GLAST = Gamma-ray Large Area Space Telescope, and C/FNOS = Communication/Navigation Outage Forecasting System.

over seven thousand that was headquartered in nearby Scottsdale, Arizona.[45] It was possible at this stage that GD would allow Spectrum Astro to maintain a degree of independence. GD subsidiaries from Gulfstream to Electric Boat had been allowed to maintain separate brand names and operational autonomy. GD was also occupied at the time with the large task of integrating land weapons businesses in Austria (Steyr), Canada (the former General Motors Canada Defense), the United States (Land Systems, purchased from Chrysler in 1982), Spain (Santa Bàrbara), and Switzerland (MOWAG).[46] Regardless of the acquisition integration path, though, one short chapter in the annals of military-industrial entrepreneurship had been closed.

That did not mean, however, that Northrop Grumman was free of small and innovative rivals. In 2004, the company was facing competition in rocket design from SpaceX, a start-up company that later took a 10 percent equity stake in SSTL. Founded by Elon Musk, a thirty-three-year-old South African who left Stanford University's physics PhD program to found Zip2 Corporation, a venture-financed software firm that he sold to Compaq Computer for $307 million. Later, he founded the Internet-payment company PayPal, which he sold to eBay for $1.5 billion. SpaceX and Northrop would end up suing each other that year over leaks of intellectual property between the two firms. SpaceX was said to have poached staff from Northrop, and Northrop was at one point overseeing SpaceX's work through a systems engineering, test and evaluation (SETA) contract at the SMC. Musk, however, had by this point acquired a reputation at least rivaling Thompson's. Lt. General Brian Arnold, commander of the SMC, told the *Wall Street Journal* at the time that Musk was a "pathfinder. We need him to be successful."[47]

Unmanned, Unafraid, and Underscoped
Success in Four Wars with the Predator
Reconnaissance-Strike Drone

The only problem with the Predator is that the Air Force didn't buy enough of them until now.[1]

Introduction

On 4 November 2002, a black Toyota Land Cruiser departed a farm near al Naqaa, Yemen, and headed down a road toward the city of Marib, about 125 miles east of the capital city of Sanaa. The departure was noted by a Yemeni undercover agent, who relayed the information to operatives of the U.S. Central Intelligence Agency (CIA). The CIA operatives quickly informed the control crew of an unmanned aircraft patrolling the area. Near a military checkpoint some distance up the road, the car veered off into the desert. Without warning, a guided missile launched from the still-undetected aircraft ripped into the vehicle. The Land Cruiser was completely blown apart, the remnants were thoroughly scorched, and the six men riding inside were left quite dead.

One of those killed, and main target of the strike, was Qaed Salim Sinyan al Harethi, whose nom de guerre was Abu Ali. A former security guard for al Qaeda leader Osama bin Laden, he was thought to have played a leading role in the attack on the USS *Cole* in October 2000, and possibly in the attack on the French oil tanker *Limburg* in October 2002. One of the dozen or so most wanted men in the world, Abu Ali had been the target of a massive manhunt. Though harbored by friendly tribesman, he had been tracked partly through his overuse of a satellite telephone. The ancestral home of the bin Laden clan, Yemen was a center of al Qaeda activity after the loss of bases in Afghanistan in late 2001. An attempt to capture al Harethi in December 2001 had left eighteen Yemeni soldiers and three villagers dead. Afterwards, the Yemeni government asked for assistance from the Pentagon, and fifty U.S. commandos were sent to the country to train Yemeni troops. The strike was said to have been personally authorized well in advance by U.S. president George W. Bush and Yemeni president Ali Abdallah Saleh.[2]

The airplane in question was an MQ-1 Predator reconnaissance and attack drone based in the former French colony of Djibouti, about 100 miles across the Gulf of Aden from Yemen. Djibouti was serving at the time as a base for over 1,000 U.S. and French commandos, and CIA Predators had been flying from its airfields for months in search of Abu Ali and others. The missile in question was a laser-guided AGM-114C Hellfire. Normally used by helicopters to attack armored vehicles and small boats, the blast from a Hellfire is sufficient to lift a truck off the ground or essentially to liquefy the occupants of a tank. Target designation came from the Predator itself, which mounted, on a stabilized gimbal, an electro-optical suite composed of an infrared camera, a daylight camera, and a modification of the Raytheon AN/AAS-44(v) laser designator used on the SH-60 Seahawk helicopter.[3]

Armed Predators are also thought to have been used to rocket the Afghan villa of Mohammed Atef, third-in-command of al Qaeda, in March 2002; to destroy the sport utility vehicle of Taliban leader Mullah Mohammed Omar; and to kill inadvertently three Afghan scrap metal collectors near Zhawar Kili in February 2002, one of whom had the misfortune of bearing a faint resemblance to Osama bin Laden. The strikes demonstrated a quite significant and innovative operational concept—a drone could quite clearly be used against pop-up targets without necessarily requiring a manned aircraft or an artillery battery for the strike. The operator, assuming that he had been granted weapons release authority, could shoot as soon as he saw a compelling target. The combination of the missile and the Predator had only been first tested in February 2001,[4] but the relative simplicity of the Predator's design and operational concept had facilitated a rapid integration. The pace of warfare had just quickened, and the sanctuary for those avoiding electronic reconnaissance and precision strike had just shrunk a little further.[5]

The Emergence of Military Drone Aircraft

Despite the novelty of the mission, the emergence of unmanned aerial vehicles (UAVs) like the Predator had been long coming. Earlier types of UAVs had flown 3,435 sorties during the Vietnam War,[6] and had been instrumental in the Israeli campaigns against Egypt in 1973[7] and Syria in 1982.[8] Over the Bekaa Valley, Israeli forces used dozens of Mastiff and Scout drones to probe Syrian air defenses. The methods were rather advanced for the time: some of the drones emitted the radar and communications signatures of fighter aircraft, enticing the Syrian defenders to activate their radars to track them, which in turn provided the Israelis considerable information about their opponents' order of battle. When the shooting started, the controllers of other drones used their laser designators to illuminate targets for precision weapons.[9]

U.S. progress before the 1990s, however, had been halting. Through the 1960s and 1970s, the Pentagon had purchased 986 unmanned aircraft, but all programs were cancelled in 1979.[10] By 1982, only 33 unmanned aircraft remained in the inventory, and all were in storage.[11] The real problem had been command and control of automated systems: long-range, autonomous operations were constrained by inadequate C3I capabilities, and this limited the number and types of missions on which the aircraft could be dispatched.[12] What changed in the early 1990s was the demonstrated availability and value of high bandwidth satellite communications, computerized mission planning and rehearsal systems, and satellite (GPS) navigation.

Indeed, unmanned aircraft have been an example of how gold-plating weapons requirements can be disastrous for practical development.[13] The Arquilla project was cancelled in the mid-1980s because development alone was expected to cost $2.17 *billion*. The rather successful Mastiff, on the other hand, had cost just $15 million to develop, so the U.S. Navy started a move toward affordable UAV acquisition in the United States by purchasing some in 1984.[14] The U.S. Army, Marine Corps, and Navy followed this acquisition with purchases of Pioneer drones (built by Mazlat, another Israeli company) over the next several years.

Developing the Predator

The impetus for an aircraft like the Predator grew out of the Clinton Administration's assumption in 1993 and 1994 that public support for its military efforts in Bosnia would collapse with the first U.S. casualties. By the early 1990s, the proliferation of Soviet-designed SA-6 missiles, as well as the early development of the Russian S-300 series, had led reconnaissance planners to assume that the U-2 Dragonlady reconnaissance aircraft would be relegated to stand-off roles.[15] Persistent reconnaissance can be a risky business, so an unmanned system offered the best way out of this conundrum.[16] The clear advantages of GPS navigation and satellite communications, as demonstrated in the 1991 Persian Gulf War, showed that UAVs could be made to fly almost indefinitely without presenting difficult command and control problems. Anticipating the value of a combined fleet of unmanned aircraft of varying capabilities, the Pentagon and the CIA collaborated on four other interrelated UAV programs, in addition to that of the Predator:

Tier I: the Gnat 750. The Predator grew out of the Gnat 750 drone that General Atomics built for the CIA in the early 1990s, which was in turn loosely based on the Amber-1 UAV that the small firm Leading Systems (see below) had built for the U.S. Defense Advanced Research Projects Agency (DARPA) in the late 1980s. One Gnat in CIA service replaced a number of two-man RG-8 Schweitzer powered gliders that were being used for clandestine reconnaissance missions over the Balkans, since the risk of losing the aircrew was quite considerable. For missions over Bosnia, the Gnat flew from an unimpressive airfield at Gjader, Albania, because the Italian government was not keen to experience the air traffic control issues of hosting clandestinely operated unmanned aircraft.[17] The CIA cancelled the operation after only roughly a dozen flights since the single aircraft performed relatively poorly in the wet, overcast weather typical of the Dalmatian winter and had problems with maintainability and datalink reliability.[18] The CIA went back in the summer of 1994 with three Gnats taken up from an order by the Turkish Air Force that Ankara, for budgetary reasons, had decided to cancel. This time they operated from Brac Island, near Split, in Croatia, in return for the supply of certain intelligence about Serbian positions in the Krajina.[19]

Tier II+: the Global Hawk. Larger and longer-ranged than the Predator is the RQ-4 Global Hawk, an unmanned aircraft with the wingspan of a Boeing 737. The Global Hawk was developed by Teledyne Ryan Aviation, builder of the Firebee and other well-known drones from the Vietnam War, as well as Charles Lindbergh's *Spirit of Saint Louis*. Ryan, notably, was bought by Northrop Grumman in 1999. The Tier II+ designation reflected the extreme range of the aircraft—in one demonstration for the Royal Australian Air Force (see below), a Global Hawk deployed across the Pacific from California to Queensland without stopping for fuel.

Tier III−: the Dark Star. Though eventually cancelled, the Dark Star program built a stealthy reconnaissance unmanned demonstration aircraft that vaguely resembled the X-45 and X-47 unmanned bombers under development by Boeing and Northrop Grumman (see also below). However, after the success and relatively low loss rates of the Predators in USAF service in the 1990s, the need for a stealthier UAV was not so apparent.

Tier III. The program for the largest of UAVs conceived by the USAF was cancelled quite early on, though some of the technology developed in the process was borrowed for the Dark Star program (which was, as noted, also cancelled). The Tier III aircraft would have been the size of a B-2 bomber—and early estimates suggested a unit cost close to $400 million.[20] Early Pentagon cost estimates are rarely underestimates (though the JDAM program to be discussed chap. 4 was a refreshing change in this regard), so the cancellation was understandable. Two of the advantages of UAVs, after all, have been their relative affordability and, thus, expendability

Nestled among these programs was Tier II: the Predator, or the "tactical endurance UAV," and an outgrowth of the design of the Gnat. Tier II was launched in 1993 as an advanced concept technology demonstration (ACTD) because many senior officials in the Pentagon had become convinced that the conventional procurement system was inefficient in demonstrating new technologies, particularly in unmanned aircraft. The ACTD is expected to field a new system in less than three years, but to do so by involving troops in the field early in the design and testing of the system.[21] The Predator ACTD, in particular, was intended to produce an aircraft that could do the following:[22]

- fly 500 miles,
- cruise at altitudes of 15,000 to 25,000 feet,
- stay on station for at least twenty-four hours,
- carry a 400 to 500 pound payload,
- provide imagery with a national imagery interpretability rating (NIIR) of 6 or better at 15,000 feet, and
- demonstrate the integration of a synthetic aperture radar with one-foot resolution at 15,000 feet.

Four potential contractors competed for the right to develop the aircraft. Three were actually teams of contractors: Teledyne Ryan and AAI, TRW and Israel Aircraft Industries (IAI), and Lear Astronics and Scaled Composites, the firm of legendary aircraft designer Burt Rutan. The fourth was General Atomics Aeronautical Systems, Inc. (GA-ASI), a firm affiliated with General Atomics, which had started life as a unit of the venerable General Dynamics but which was better known as a builder of nuclear reactor control systems. GA had only entered the aircraft business in 1993 by buying Leading Systems, the roughly ten-person firm that had built the Gnat; in the process, GA effectively rescued the company from bankruptcy.[23] GA quickly moved to professionalize its processes. The new aircraft group was soon headed by retired Rear Admiral Thomas Cassidy, who had once headed the U.S. Navy's Top Gun Fighter Weapons School at the Miramar air station near San Diego.

In January 1994, GA won the roughly $32 million contract for the ACTD.[24] In the first phase of its work, GA was expected to deliver three reconnaissance aircraft and a ground control station in just six months, and demonstrate that the aircraft could be flown with an on-board infrared camera and an ultra-high-frequency (UHF) satellite communications link. Attempts to complicate the program with add-on requirements were, in the words of one Pentagon official, "successfully stiff-armed."[25] General Atomics met the deadline five days early with a twenty-second flight in July 1994. The company's first contract from the Pentagon for ten aircraft was awarded later that year. Over the next two years, Predators were evaluated in exercises and on operations.[26] Within just one year, the Predators were flying operationally from Gjader, Albania.

The Advantages of the Predator and Its Technology

Planned from the start as a reconnaissance aircraft, the Predator's basic sensor package today includes a television camera, a Versatron Skyball Model 18 FLIR, and a Westinghouse 783Z234 synthetic aperture radar (or SAR, which was actually not available in 1995). These components were off-the-shelf items modified for the form of the airframe and dynamically recombined into an integrated reconnaissance system. The Predator is otherwise a relatively simple aircraft, at least by military standards. The initial version of the Predator has a wingspan of 48 feet, is only 26 feet long, and weighs but 1,500 pounds fully fueled. The original engine, which put out only 80 horsepower, was a modification of the Rotax 912-piston engine originally designed for snowmobiles. Though the engine only pushed the aircraft to about 80 knots, the aircraft was not so vulnerable at high altitudes: the small fiberglass airframe with the Kevlar skin was rarely successfully tracked by USAF radars in testing.[27] In any case, losing one was no great loss: the aircraft and its full sensor package were priced at about $3.2 million each.[28] Keeping one airborne, though, can be profitable: the Predator can stay aloft for up to forty hours and routinely performs twenty-four-hour missions in which the control crews change at least once. This allows the aircraft to operate effectively up to 1,000 kilometers from its base, transmitting video and radar imagery to ground stations almost anywhere in the world.

For operations over Kosovo, a targeting laser was added for guiding bombs (see below). By the campaign in Afghanistan, Predators had begun to fly as reconnaissance-*strike* aircraft with two Hellfire missiles—the first test of a Predator with Hellfire having come in February 2001.[29] General John Jumper, the USAF chief of staff, had reportedly wanted to mount weapons on Predators for some time,[30] but General Atomics had earlier claimed that converting the drone into a strike aircraft would be quite difficult. This was apparently done in the hopes of persuading the State Department to approve export licenses that could have been more difficult to obtain for a *combat* aircraft.[31] This evolution, however, was not the last for the Predator. The aircraft was later developed into the turboprop-powered MQ-9 Predator B, or Reaper—an aircraft that could carry fourteen Hellfires. The Predator may even eventually transport on operational flights its own UAV, a fifty-eight-pound expendable aircraft called the Finder, which would drop down to lower altitudes to check for toxic agents.[32] In June 2002, a Predator launched one in a test over Edwards Air Force Base.[33]

Predator Operations

The potential of the aircraft was not lost on the Pentagon, and during the next four significant campaigns in which U.S. forces fought, Predator units distinguished themselves:

In early July 1995, Predators under the control of Air Force Material Command and Army Intelligence deployed to Albania and Croatia to monitor Serbian troop movements and positions. During the campaign in August and September, the aircraft flew fifteen missions, undertaking reconnaissance and battle damage assessment for aircraft and cruise missile strikes, and picking out individuals from a range of two to three miles.[34] Since most of the flying was at night from relatively expeditionary airfields, the crews found themselves trying to sleep during the hot summer days in tents.[35] Two aircraft were lost. On 11 August, a Predator equipped with a new Ku-band video satellite communications link dropped below 4,000 feet in altitude to investigate a Serbian vehicle column, and a nearby anti-aircraft gun brought it down. On 14 August, another drone experienced a completely unknown engine fault and began dropping from its operating height of 20,000 feet. The damage could have been from shrapnel from an anti-aircraft missile or gun, as the remote telemetry indicated a completely unanticipated set of indications. After the aircraft had been dropping toward the coast for about thirty minutes, the crew determined that it would not make the water, and ditched the drone into the side of a mountain. Three aircraft returned to the airspace of the former Yugoslavia in 1996, flying out of Taszar, Hungary, on surveillance missions with the new synthetic aperture radar, and typically performing one ten-hour mission daily, six days each week.[36]

In 1999, nineteen RQ-1A Predators of the USAF's 11th Reconnaissance Squadron logged over 2,000 flight hours flying from Tuzla, Bosnia, and Tirana, Albania. The aircraft were mostly used for monitoring areas where Yugoslav troops were thought to be operating and relaying targeting and bomb damage assessment via satellite to intelligence analysts. The Predator's sensors did have difficulty with the expert concealment tactics of the Yugoslav Army and the adverse weather experienced that spring over Kosovo.[37] Three of the nineteen, however, were used for designating targets with lasers, and toward the end of the campaign, plans were being laid to arm them with Hellfires. The leap from equipping the aircraft with laser designators to equipping them with laser-guided missiles was a logical one.[38]

Five other types of NATO drones flew in the war. Twelve U.S. Army RQ-5A Hunters logged about 900 hours reconnoitering Yugoslav troop positions. Five U.S. Navy RQ-2Q Pioneers flew from USS *Ponce de Leon* (a troop-landing ship with a large helicopter pad) to keep tabs on the Yugoslav Navy. [The Pioneers had earlier distinguished themselves in the 1991 campaign against Iraq.] British Phoenixes of the 32nd Regiment of the Royal Artillery flew twenty sorties looking for targets for NATO aircraft. French and German CL-289s Pivers (*Progammation et Interpretation des Vols d'Engins de Reconnaissance*) flew about 180 sorties,[39] and French Crecerelle drones flew as well. In total, twenty-one unmanned aircraft were lost to a combination of accidents and air defenses.[40] After the war, drones proved rather valuable in monitoring the province for riotous activity. As Captain Jon Harding, operations officer for 57 Battery, Royal Artillery, put it, "people won't leave houses when Phoenix is overhead—they think they are being watched."[41]

AFGHANISTAN

Immediately after the Kosovo campaign, the newer and much larger MQ-4 Global Hawk became available for limited operations. By June 2002, Predators and Global Hawks of the USAF had logged 1,000 combat flight hours over Afghanistan in the campaign against the Taliban and al Qaeda.[42] Two Predators and a Global Hawk were lost in the first several months of fighting and occupation, through both apparently due to adverse weather, not ground fire.[43] The Global Hawks were initially based at Ramstein air base in Germany, flying to Afghanistan on a six-hour route over Austria, Hungary, Romania, the Black Sea, Georgia, Azerbaijan, the Caspian Sea, and Turkmenistan.[44]

IRAQ

Predators were used on a wide variety of missions, from monitoring troop movements around Baghdad to patrolling the western deserts of Iraq for attempts to launch ballistic missiles at Coalition forces or Israeli cities. Before the ground fighting began, one Predator crew attempted to dogfight an Iraqi Air Force MiG-25 using the aircraft's Stinger anti-aircraft missile, but unsurprisingly, lost the engagement.[45] The USAF lost four Predators, but two of these losses came when the controllers deliberately flew them until they ran out of fuel—the aircraft were thought at those moments to be too valuable as a means of drawing Iraqi fire to return them to base.[46]

More generally, at least ten different types of drones were used by U.S. forces in the war against the Ba'athists in Iraq in 2003. There were also reports that the USAF operated a secret and stealthy drone, built by Lockheed Martin and based on the cancelled Dark Star aircraft, from Al Ubeid air base in Qatar.[47] Allied forces used a few more types in addition, but the Predator as well.[48] The British Army lost over Iraq some twenty-three of the 198 Phoenix drone aircraft it had acquired since 1996. The apparent general vulnerability of the unmanned aircraft led the Ministry of Defense to reevaluate its plans to acquire another developmental UAV and to consider instead purchasing a less costly, proven, off-the-shelf design in the coming Watchkeeper program.[49] It also led, as an interim measure, to the establishment of an RAF Predator unit at Nellis Air Force Base. Commanded by an Royal Air Force (RAF) squadron leader, 1115 Flight was composed of 45 troops, mostly from the RAF, but including a few USAF airmen as well. The operations of this unique, combined, tactical unit allowed the RAF to "kick the tires" on the Predator before buying its own armed drones.[50] The *Aeronautica Militare Italiana* also took that step, taking delivery of four MQ-1B Predators in January 2006. One of the aircraft crashed in Iraq that May but was later repaired and returned to service.[51]

Even after the Iraqi Army and Republican Guard were defeated, the Predators remained in Iraq to support the ongoing counter-insurgency campaign. Organized as the 46th Expeditionary Reconnaissance Squadron, the Predators and their support crews operated from Balad airfield, flying over 100 missions per month. A testament to the economy of the Predator concept was the size of the unit: the 46th was organizationally a flight, led by a major (effectively, a *squadron leader*) or a captain, and was staffed by only fifty-five troops (see below). At the same time, it launched, recovered, and serviced roughly twenty aircraft at a time.[52] Despite the low manning, Predators eventually became almost the preferred method of fire support in the ensuing fighting. During the Second Battle of Fallujah in November 2004, Predators were repeatedly used to kill individual insurgent snipers on rooftops. The battle even demonstrated the ability of dissimilar UAV units to cooperate with one another. U.S. Marine Corps (USMC) Pioneer UAV crews based in Iraq used real-time electronic chat to communicate with USAF Predator (combat) crews 8,000 miles away in Nevada. At one point, a Pioneer crew that had tired of attempting to kill an insurgent mortar team by calling in 155-mm shells (the mortarmen had timed the shells well, ducking for cover as they arrived) instead resolved "we're getting a Predator" and took down the entire position with a Hellfire.[53]

In these campaigns, the Predator and drones like it captured public imagination largely because of their unusual combination of novelty and retrospective: while piloted remotely, they are propeller-driven aircraft that fly at speeds and altitudes akin to those of a Sopwith Camel.[54] Their cost is reasonable, and their units are, in several ways, rather modest affairs. A basic Predator flight consists of four aircraft, a ground control facility, an operations center, a satellite communications van, and as noted, fifty-five officers and ground crew. About five flights comprise a squadron, and a single flight can maintain twenty-four-hour surveillance over its designated area of responsibility. Operations cost about $100 per hour, exclusive of the high cost of satellite bandwidth; sorties of manned tactical aircraft, on the other hand, cost at least $1500 per hour. By 2006, the 11th, 15th, and 17th Reconnaissance Squadrons had roughly fifteen flights between them with over 1,000 officers and troops. National Guard squadrons from Texas, Arizona, California, and New York have received aircraft as well and contribute several hundred more airmen each. All in all, the numbers represent quite a modest force for the capability it represents.[55] The USAF even had tentative plans to expand the force to fully fifteen squadrons, and General Atomics kept on expanding its production capacity to keep up with the demand.[56] As USAF Chief of Staff General John Jumper told a Congressional committee in February 2005, "we're going to tell General Atomics to build every Predator they can possibly build."[57] How the USAF would control all those aircraft remotely, however, was another question, to which I now turn.

Problems with the Predator

While the Predator is one of the best recent examples of the potential for spiral development efforts, this is not to say that the aircraft is without issues. Before the Afghan campaign, the plane failed the rather strict operational testing regime to which the USAF subjects developmental aircraft. The Predator, it was known, could not take off in heavy rain, snow, ice, or fog. Its low speed left it rather vulnerable to ground fire, at least at the lower altitudes at which it did its best work. Its sensors, that is, work best at altitudes under 10,000 feet, which puts it easily within the range of anti-aircraft cannons. While the control stations cost only about $12 million each, they are rather awkwardly designed.[58] Fortunately, the Pentagon did not allow the bureaucratic hurdle of a negative test report to dampen its enthusiasm for the concept. More so, the design of the Predator B ameliorated many of these problems, increasing the aircraft's top speed from 120 to 220 knots, its payload from 450 to 750 pounds, and its ceiling from 25,000 to 45,000 feet. In return, the wingspan would increase by about 16 feet and the length by about 7 feet.

A particular concern about the use of armed drones is found in the potential for collateral damage. In general, removing the pilot from the aircraft
arguably decreases his situational awareness—the cockpit provides considerably greater immediacy than a comfortable chair and a computer monitor
hundreds or thousands of miles from the battle. Pilots in the cockpit may
be better at distinguishing between combatants and noncombatants on the
ground. Even if this is just a perception by politicians and citizens, it could be
an important one.[59] Of course, despite the aforementioned incident with the
scrap traders, a considerably larger number of Afghan civilians were killed in
July of the same year when a USAF AC-130 gunship opened up with its machine guns on their wedding party after mistaking their *feu de joi* as hostile.[60]
Since an AC-130 carries a crew of up to fourteen, it is arguable that even heavily manning the aircraft is no prophylaxis against noncombatant casualties.
On the other hand, examples can be found of situations in which the use of
unmanned aircraft might actually have averted unintentional friendly casualties: the two Illinois Air National Guard F-16 pilots who killed four soldiers
of Princess Patricia's Canadian Light Infantry Regiment in April 2002 were at
the end of a six-hour patrol, and were under the influence of amphetamines
prescribed specifically to keep them awake in the cockpit.[61]

A further, and huge, problem is the scarcity of satellite radio bandwidth.
Even though the USAF is planning to maintain a fleet of over one hundred
Predators,[62] the USAF will hardly be able to fly them all simultaneously. Plans
for large numbers of UAVs were laid in the late 1990s when huge private
investments in worldwide telecommunications infrastructure suggested that
up to 1,000 communications satellites might be available within ten years to
carry all the required communications traffic. Piggy-backing a few military
videophone calls on that infrastructure would have been easy. Instead, the
commercial telecommunications satellite business took a downturn as sharp
as that of the global fiber-optic cable industry. Between 1998 and 2002, the
pace of commercial launches was just over one-third what had been expected.
Thus, in 2001 and 2002, the USAF was able to keep only two Predators and
one Global Hawk operational over Afghanistan simultaneously.[63] One partial
solution may be to load these aircraft with terabytes of information on the
areas they are intended to surveil so that "they only need to send a message if
somewhere in the image a pixel changes."[64]

The Predator, of course, is not quite a fighter-bomber. It cannot refuel in
flight. It cannot fly in formation. General Hal Hornburg, commander of the
USAF's Air Combat Command, asserted in late 2003 that he considered these
two issues, in particular, to be major stumbling blocks for the advancement
of unmanned military aviation.[65] On the other hand, altering the concept of

operations can get around these problems, so the necessary solutions may not be completely technological.

Adapting to Innovation

Integrating unmanned aircraft into a force structure designed for manned aircraft should be expected to pose some challenges. For the U.S. Navy or the USMC, there were operational and cultural questions to answer, but neither service is fundamentally built around aviation. For the USAF, this was a different matter. The USAF has assigned only commissioned aviators with previous cockpit experience to fly the Predator and Global Hawk, and the USAF has so far been adamant that this practice continue. To do otherwise would seriously change the culture of the service. Despite its claims to be an air and space force, and over forty years experience with missile forces, no one but a career pilot has ever held the office of chief of staff in the USAF.[66]

Pilots of manned aircraft, however, have generally not been thrilled with being assigned to drone squadrons. Very few of these "flyers" are volunteers, as most have expressed a preference for flying in the cockpit, rather than remotely.[67] A few of the pilots who have been ordered to Predator duty have chosen, if their required service was coming short, to refuse the assignment and leave the service instead. Faced with this problem, an air arm can consider at least two options, one within the context of the silk scarf culture and one rather outside it. The USAF has been crediting UAV hours toward pilots' flight pay determination to improve both compensation and morale. Without this, their pay and their chances for promotion would suffer.[68] The service has also been supplying chase aircraft to Predator squadrons to keep pilots' in-the-cockpit skills relatively up-to-date. Alternatively, if fighter pilots are naturally disinterested in controlling drones, the drones can be sent to a different corps of professionals. At one point, the USAF was considering allowing a few enlisted airmen who were FAA-certified private pilots to fly the aircraft from their remote control stations—a number had sent unsolicited resumes to the 11th Reconnaissance Squadron in 1995.[69] In current operations, the person controlling the Predator's mission is more often the enlisted sensor operator, who often finds himself in the unique position of giving orders to the commissioned pilot sitting alongside him.[70] This may not be surprising. In the U.S. Navy, Army, and Marine Corps, pilots for certain types of drones are often drawn from the enlisted ranks. In the Vietnam War, the pilots for Firebee drones were often contractors from Ryan Aeronautical.

Many other services have dedicated drone pilots; for military services not dominated by pilots and their aircraft, this has been easier. In the British,

French, and German armies, truck-mobile drones (such as the Phoenix and the Sperwer) have generally been assigned to the artillery. There are at least two reasons for this. First, the drone units operate much like artillery forces: following the troops in the field, operating in small groups, and launching strikes (or reconnaissance missions) upon short-notice orders from the front lines. Without a fixed-base structure, extensive cockpit schooling, and a silk-scarf culture, drone units seem to many to fit better here than alongside manned aviation.

Some innovations, that is, are easier to digest organizationally than others. When a UAV is used to fly a surveillance patrol over a downed pilot awaiting rescue—when the risk of losing another pilot is not thought justifiable until a large force can be assembled for the mission—few will agitate about the importance of a person inside the aircraft. On the other hand, getting UAVs ready for close air support is an entirely different matter in material, cultural, and doctrinal terms.[71] While this is neither surprising nor unfortunate, UAV adoption in the USAF may have proceeded more slowly than necessary due to an institutional bias against unmanned aircraft in the service.[72]

The Pace of Innovation

Unmanned aviation, at least in the USAF, had reached this juncture largely because of the pace of innovations in this sector since the mid-1990s. As one USAF officer put it, "one of the beauties of the UAV . . . is that the developmental timeline to integrate new technology seems to be significantly reduced."[73] Crashing an aircraft does not tend to kill the pilot, so risks can be more readily taken in product development. UAVs certainly have maintained a reputation for expendability. Of the one hundred Predators produced through February 2004,[74] roughly one-third had been lost in accidents or combat.[75] Naturally, loss rates like this can be a problem on campaign if resupply is not swift and spares are not plentiful. On one of its deployments to Afghanistan, the Canadian Army ran out of its Sperwer drones after the sixth and final one available crashed. The six had flown only twenty-nine missions, and the loss left the Canadians relying solely on the Germans' Luna UAV for surveillance and reconnaissance.[76]

While a few have criticized this as wasteful of resources, the accompanying renewal of the asset base has spurred innovation to a degree unseen in other sectors of the aircraft industry. Innovation in transport aircraft has been continuous, but there are limits to what can be achieved with new designs under the existing state of technology. Boeing has been touting its 787 Dreamliner for the 7–8 percent fuel savings that users may experience. For airlines, this is real money; for military services, it is useful but does not represent a quantum leap in performance. Airbus's A380 is an engineering marvel, but primarily for its heretofore unseen passenger capacity. It is thus no wonder that Airbus and Boeing compete in the transport business on price as much as technology. The rotorcraft business has experienced a similar dynamic. Eurocopter has captured the leading share of the civil helicopter market with a long series of incremental improvements. Tiltrotors present a potential great leap forward, but only if the technology can be demonstrated to work safely with consistency. In precision weapons, the addition of laser guidance in the 1960s and satellite guidance in the early 1990s were discontinuous improvements, but sudden ones. After the introduction of these technologies, the state of the art reached a relative plateau. GPS-guided weapons have experienced decreasing circular errors probable (CEPs) over the past decade, but largely because of improvements to the satellite and control segments of the Navstar system. Thus, there should be no surprise that market shares have remained relatively stable in this sector as well.

One could be excused for surprise at the relative success of the Predator program, if only because the track record of U.S. military services in procuring developmental drone aircraft had been rather poor.[77] The Army, as noted, spent years and billions of dollars on its Aquilla remotely piloted vehicle program in the 1980s, and ultimately produced little more than some interesting technologies. The Navy's off-the-shelf purchase of Pioneer UAVs from Israel Aircraft Industries (IAI) and AAI in the same decade was much more successful. The aircraft were purchased to meet a very specific deficiency. When the battleship *New Jersey* shelled Syrian air defense positions in the Bekaa Valley in February 1984, it did so without any aerial reconnaissance or battle damage assessment capability—naval aviation had already been bloodied by the loss of two A-6 Intruder jets. The Pioneer had many operational limitations, but it flew and worked relatively well, and the Navy resisted the requirement creep that could have doomed the program. By the time of the first war in Iraq, the Pioneer type was in heavy use. In February 1991, a large party of Iraqi troops attempted to surrender to a Pioneer overflying their positions on Fayluka Island, Kuwait.

This U.S. experience with UAVs also demonstrates how military transformations are sometimes not so revolutionary, at least not immediately. Creating new ways of warfare is not always an all-or-nothing proposition that requires scrapping all the old weapons and buying only new ones.[78] The last combat use of U.S. battleships came in 1991, not in 1945. In the 1991 campaign, the battleships *Missouri* and *Wisconsin* were used to launch both cruise missiles and Pioneer reconnaissance drones against Iraqi targets. Today, the relatively low unit cost and remote operability of the Predator and other drones allows for much greater operational experimentation than do manned aircraft, and experimentation is essential to transformation.[79]

Specialization Is an Option

General Atomics' success in the unmanned aircraft field is a clear indication that scale and scope are not sufficient or necessary for success in every sector of the arms business. Scale and scope are frequently touted as essential in assembling the systems of warfighting systems that defense ministries increasingly seek. Today, however, many of these capabilities are available to considerably smaller contractors who have the skills needed to integrate the components into platforms or larger systems. General Atomics produces the Predator's airframe, but the components essential to its primary mission of reconnaissance are all procured outside the company. As noted, the aircraft's synthetic aperture radar came from Westinghouse, its visual sensors from Wescam, and its laser designation system (which proved essential over Afghanistan) from Raytheon—a rather large mission systems integrator in its own right. General Atomics' latest version of its flagship product, the Predator B, is to be a turbo-prop, hunter-killer attack aircraft with a ceiling of 50,000 feet that can carry up to eight Hellfire missiles[80] or the 500-pound GBU-38 version of the Joint Direct Attack Munition (JDAM).[81] In these cases, the weapons will come from Raytheon and Boeing. Taken together, these subsystems can conceivably close the sensor-to-shooter gap against mobile weapons of mass destruction (the Pentagon's holy grail) to seconds, a task on which the large systems integrators have been working for years.[82] Simply put, General Atomics is today itself a systems integrator of relatively small scale but wide scope, and the owners are pursuing greater systems integration capabilities. While these capabilities do not provide the company the weight needed to pursue the largest defense programs, they do make it extremely competitive for contracts that are not vast.

General Atomics does face considerable competition. In the unmanned aircraft business, the market leader is probably Israel Aircraft Industries (IAI), with over $250 million in annual unmanned aircraft sales, or a 25 percent share of the global market.[83] IAI's success is partly the result of what Michael Porter has termed a selective factor disadvantage.[84] Since the Israelis have been known to trade hundreds of Arab guerrillas and bandits for a handful of pilots, limiting the exposure of manpower in a nation of less than five million people is particularly important. That the Israelis need to economize on manpower is a selective disadvantage that leads to an industrial advantage in drone innovation and furious and focused interest in unmanned aircraft.[85] Geographic clustering may also have been a relative Israeli advantage, as collocation of research, design, production, and testing has been shown to be quite valuable in the aircraft business.[86] One of the salient problems that must be tackled in unmanned aircraft development is the replacement of the pilot in the cockpit with an electronic system capable of controlling the plane from a distance, or autonomously if possible. The Israeli software industry, it is well known, is in part a creation of the Israeli military's need for advanced weaponry. The MAMRAM computer science institute and its School for Computer Related Professions have supplied a large network of programmers and computer consultants who form a rather dense network of computer systems knowledge in a country the size of Connecticut.[87] This is fortunate, as the relative surprise of facing UAVs flown by Hezbollah in the 2006 campaign in Lebanon was further evidence that Israel's arms industry needs to strive continuously to stay ahead of enemy developments.[88]

With its ACTD, General Atomics was relatively free from the modern U.S. burden of producing weapons systems components in fifty different states. The Pentagon's Joint Unmanned Combat Air System (J-UCAS) program was managed along similar lines: the two competitors were developing their entries at relatively self-contained facilities at the USAF's test center at Edwards Air Force Base in southern California. Many of the Boeing engineers who worked on its unsuccessful entry in the Joint Strike Fighter competition later worked on the company's X-45 J-UCAS demonstrator, while Northrop Grumman had been teamed with Lockheed Martin in its work on its X-47 demonstrator. Each promised to be a relatively stealthy, unmanned strike aircraft with an airframe built from nearly 90 percent composite materials.[89] Both programs were cancelled in 2005, largely due to the continuing success of the Predator. The United States did not arguably need another land-based drone with a similar mission and endurance, albeit with greater speed and payload. What it could use was a carrier-capable drone that could provide persistent, armed overwatch of areas reachable only by carrier-based jets—such as the interior of China. Thus, in late 2005, the UCAS mandate was sent back to the USAF and the Navy individually for separate development of land- and carrier-based drones, as appropriate.

European firms are also chasing similarly novel concepts in combat aircraft. While BAE and the British Ministry of Defence have been exchanging stealth aircraft design information with U.S. firms and the Pentagon, the German, French, and Swedish defense ministries have been conducting information-sharing of their own. Many forget that stealth aircraft research actually started in Germany in 1975, which produced the Lampyridae demonstrator in the 1980s and a number of other demonstrators in the 1990s. The X-31 program, funded by the U.S. Navy and the Luftwaffe, and undertaken by Boeing and EADS, aimed to produce a short takeoff and landing fighter aircraft with thrust vectoring—at significantly less cost and complication than Lockheed Martin's award-winning lift fan (which has recently experienced some developmental difficulties in the Joint Strike Fighter program). Today, much of EADS's stealth efforts are aimed at producing penetrating UAVs for suppressing air defenses.[90]

Further down market, a wide range of small firms in less-developed countries have unmanned aircraft programs underway. The minimum efficient scale of development and production for unmanned aircraft is considerably less than for manned aircraft, and the lure of a secure source of reconnaissance (or even strike) aircraft is a strong motivation to invest in a relatively small development program. Even the Iraqis, with their limited industrial base in the 1990s, built UAVs, even if reports that they had planned to use them for chemical attacks proved unfounded.[91]

Small Firms and Small Aircraft: The Predator as a Case Study

Reviewing the criteria laid out in chapter 1, it is evident that General Atomics' success with the Predator program was no fluke. The aircraft exemplifies the type of system in which small arms contractors can be quite successful vis-à-vis their larger competitors.

> *Innovation without R&D intensity.* Unmanned aircraft were not a new concept in 1994, but the type of unmanned aircraft that the Predator became certainly was. At the same time, the reason that the aircraft was assembled in just six months was that it was based on a solid design (the Gnat 750)—of which but a few examples had been built—and its own evolved design was based on the recombination of off-the-shelf technologies. The market was also developing rapidly. After fewer than one hundred aircraft, GA-ASI had moved on to a turboprop-powered version with far heavier ordnance and a related maritime version for over-water patrols.

> *Skill-intensive production.* Sources at Northrop Grumman have suggested that a Global Hawk production rate of two per month would be optimum.[92] General Atomics was producing several Predators per month, but the company was a bit secretive about just how many. Regardless of the details, this rough rate of production suggested that assembly was not as machine-intensive as automotive production. Indeed, in the military-industrial realm, it was closer in skill-intensity to shipbuilding than guided missile manufacturing. In capital intensity, it falls considerably below shipbuilding, in that most of the assembly

work is accomplished by hand—the final product weighs about one ton, not 1,000 tons (see chap. 5).

Medium-speed learning. This was again true of the unmanned military aircraft business in the 1990s: a wide variety of aircraft were ordered by a few dozen military services around the world, but none of the aircraft were produced in great quantities. Some were built by small firms like GA-ASI, and others were the work of giant firms like EADS. At the time, the largest aircraft companies in the world were concentrating managerially on merging with one another and on executing a handful of large and complicated manned aircraft programs. This provided considerable opportunity for an alacritous competitor to gain a foothold in a niche market. Over the next ten years, GA-ASI built only about a hundred Predators—which suggests a rate of learning-by-doing not exactly that found in automobiles but better than that found with space shuttles.

In light of this evidence, it should be no surprise that relatively small firms like GA-ASI and IAI have taken leading positions in the unmanned aircraft industry. By 2002, General Atomics had 650 people working on unmanned aircraft projects, and by 2007, it had over 2,000. Throughout this time, it was one of the fastest-growing arms contractors in the U.S.[93] That did not mean, however, that General Atomics could do everything alone.

Enter Broad Area Maritime Surveillance and Lockheed Martin

The previously mentioned Predator B or MQ-9 Reaper, was developed in 2000, and took its first flight in February 2001 with its 700-horsepower rear-mounted Honeywell TPE-331-10 turboprop engine, driving a three-blade propeller. Shortly thereafter, General Atomics was pitching a maritime version of the Predator B—the Mariner—to the U.S. Navy to fulfill its Broad Area Maritime Surveillance (BAMS) requirement. The Navy would base squadrons of the aircraft at five bases around the globe in order to cover as much of the world's oceans as possible from a minimal number of home sites: at Naval Air Station (NAS) Kadena in Japan, on Diego Garcia, at NAS Sigonella in Sicily, at NAS Jacksonville in Florida, and at the USMC base on Kaneohe Bay on Oahu. One advantage of a UAV in this circumstance is that close surveillance of a hostile coast does not risk the loss or capture of an aircrew, as when the Chinese Air Force seized a Navy EP-3 Aires aircraft in the spring of 2001.[94] In May 2004, the U.S. Navy and Coast Guard were shown what the Mariner could do off San Diego—this was the first time that an unmanned U.S. aircraft demonstrated 360° radar surveillance at sea (the table shows the details of the Predator B family).

CHARACTERISTICS OF GENERAL ATOMICS'S PREDATOR B AND DERIVATIVES

Aspect	Predator B Surveillance and strike	Altair Scientific research	Mariner Maritime surveillance
Wingspan	66 feet	86 feet	86 feet
Fuselage length	36 feet	36 feet	36 feet
Weight	10,000 pounds	7,000 pounds	10,500 pounds
Ceiling	50,000 feet	52,000 feet	52,000 feet
Endurance	30 hours	30 hours	49 hours
Internal payload	800 pounds	660 pounds	800 pounds
External payload	3,000 pounds	3,000 pounds	3,000 pounds
Air speed	220 knots	TBA	220 knots
Primary customer	USAF	NASA	U.S. Navy (intended)

General Atomics was not this time the sole prime contractor for the Mariner. Rather, the company had invited industry heavyweight Lockheed Martin to team up with it to produce the Mariner and its C4ISR package. Indeed, the latter was largely integrated by Lockheed and included such components as a Lockheed Martin AN/UYQ-70 Valiant display workstation, a next-generation variant of the Q-70 displays, controls, and operating systems that were widely used in Navy applications on sea, land, and airborne military platforms.[95] With the upgraded sensor package, including its Raytheon SeaVue multimode maritime and overland surveillance and synthetic aperture mapping radar, performance would be impressive. The aircraft would be able to detect a man on a raft at 30 nautical miles (nm) and a supertanker at 230 nm.

The Navy had planned to select a single contractor to continue development in the spring of 2005, and planned to buy two aircraft in 2007 and four each in 2008 and 2009.[96] The USAF, however, asked the Navy to award a sole source contract to Northrop Grumman so that both services could operate the Global Hawk. As the larger aircraft (including its sensor package) cost (in unit terms) $35 million each, roughly three times as much as the admittedly shorter-ranged Reaper, the Navy declined.[97] Still, the Australian Defence Ministry had announced earlier that year that the Royal Australian Air Force would procure roughly ten Global Hawks for long-range maritime and overland reconnaissance, so the idea had some precedent. Such a purchase would have moved the Global Hawk far further down its cost curve than a greater purchase of Predators, which were already being bought in some quantity. It also would provide impetus to the unmanned aviation program of a very large firm. On the other hand, General Atomics won another important contract in Australia. National Air Support, the firm that supplies fixed-wing surveillance aircraft to Australian Coastwatch, a branch of the Australian Customs Service, turned to the San Diego firm to replace its mixed fleet of fifteen Islanders, Cessna F-406s, and De Havilland Dash-8s.[98] This was an obvious choice, as there was little reason to pay aircrews for purely surveillance missions. The U.S. Navy, however, remained disinclined to participate in the USAF's plans; interservice jockeying (and a host of other reasons) then led the Navy to postpone the competition until at least 2008.

The Question of Alliances

In the BAMS competition, General Dynamics was also offering an unmanned version of its Gulfstream V light transport jet, which had lost the interrupted Airborne Common Sensor contract to Lockheed Martin's entry of an Embraer ERJ 145, which was itself another case of Lockheed Martin teaming with a smaller aircraft manufacturer that possessed an enviable command of its segment of the business. This sort of teaming arrangement was successfully used by Thales of France and Elbit of Israel in offering a Thales-outfitted fleet of Elibt-produced Hermes drones to win the British Army's Watchkeeper UAV competition.[99] This suggests that alliances may play a valuable role in this sector. Reviewing the criteria laid out in chapter 1, one finds reason for optimism:

> *Considerable change with respect to processes and goals.* Unmanned aircraft, without question, constitute a technologically preparadigmatic sector of the military aerospace industry. Ask a group of engineers or fighter pilots to describe what the next fighter aircraft should resemble, and the list of responses will cluster around some definable parameters. This is not so with drones: the most numerous types in the U.S. inventory—the Hunter, Pioneer, Predator, and Global Hawk—all conduct reconnaissance but differ considerably in form. The same is true with respect to operational roles: debate continues as to whether unmanned systems should serve as aerial tankers, penetrating bombers, close air support aircraft, or interceptors. With so little settled about the industry, mid-term alliances can preserve firms' options for future projects without locking them into technological choices that are too narrowing too quickly.
>
> *Moderately leaky knowledge.* The Lockheed Martin–General Atomics alliance shared an important characteristic with the Thales-Elbit alliance: in both cases,

the designated prime contractor was a systems integrator that needed to look outside the firm for specific expertise in unmanned aircraft. Airframe design and flight control software would be found in its alliance partner, and the larger firm would contribute C4ISR systems acumen. These two levels of integration—the platform and the payload, respectively—would interact considerably, but they would not be so closely integrated that critical intellectual properties would slip away without warning. On the other hand, the payloads could not simply be bolted onto the platforms the way a machine gun can be bolted onto the top of an armored truck (see chap. 7).

Moderate potential for a shakedown. By signing on with the larger firm, GA-ASI was admitting that it lacked some integration skills and marketing clout that could be found in a Lockheed Martin. At the same time, its Predator and Mariner aircraft could, after a period of disengagement and some switching costs, be marketed through another systems integrator or on GA's behalf itself.

A Slow Shakeout and Millions at Stake

For GA, alliance strategy was not an all-or-nothing matter. In August 2005, GA was chosen by the U.S. Army for its Extended Range/Multi-Purpose (ER/MP) drone program. The aircraft in the bid was the MQ-1C *Sky Warrior*—and it was entirely a derivative of the Predator.[100] The Warrior was designed as a corps commander's hip-pocket air arm, with a squadron attached to each such formation for long-range reconnaissance and attack. The aircraft would have a shorter range, but stronger wings for carrying up to fourteen Hellfires. Since the aircraft was so similar in purpose to the Predator, GA needed no partner to capture the contract. The similarity, however, did draw the attention of Pentagon procurement chief Ken Krieg, who insisted that the Army and Air Force sort out their respective drone requirements before proceeding.[101]

In the BAMS competition, at least, the alliance with Lockheed Martin seemed very sensible for GA, though with one caveat and one remaining question. The caveat was that even the largest and most seemingly accomplished allies could falter. Lockheed Martin actually subsequently lost its contract to design and build the ACS after the Army learned that the ERJ 145 would be too small to accommodate all the sensors specified. Though a storied aircraft builder, Lockheed had assigned only engineers from its electronics divisions to the project. Sound alliances are not formed instantly upon contract signings: some degree of knowledge transfer is vital.[102] The remaining question is whether firms like GA-ASI and Elbit will continue to hold the leads that they have developed. The unmanned aircraft market today may be rather like the commercial aircraft market of the 1920s and 1930s. Transport aircraft had just become commercially useful, and the market was expanding briskly. Designs had not become paradigmatic, so clever engineering could easily beat large investments in systems development. However, by the 1930s, the industry had started to consolidate. This was in part due to the federal legislation that showered government favor in the form of airmail cargo contracts on larger, more established airlines. This, in turn, led to larger purchases from larger aircraft manufacturers, which may have prematurely consolidated the market. The greater reason, though, was the rapid development of technology—mistakes made in launching new products with very high development costs could be fatal to some of the smaller companies involved.[103]

Unmanned aircraft may not face such a brutally Darwinian dynamic in the short run, largely because drone aircraft research is less expensive and drone aircraft testing is more accessible. Greater risks can be taken because, as noted, the unit costs are rather low and there are no test pilots to kill. GA demonstrated in the 1990s that a relatively small contractor could effectively integrate modern mission systems on a single platform. Half the market for UAVs, however, lies not in airframes but in their payloads and ground stations, and the airframes constitute necessarily a smaller portion of the expense than with manned aircraft.[104] Larger contractors, like Boeing and Northrop Grumman, have been loudly telling investors, procurement officials, and industrial policy-makers that only large firms (like them) can master the integration skills needed to develop systems of systems for network-centric warfighting among platforms. Indeed, there is a great deal of money in this market. As Eric Hansen of the U.S. Naval Surface Warfare Center in Carderock, Maryland, put it, the development of command-and-control systems that can operate any type of unmanned vehicle and permit uninhabited sharing of information collected from various unmanned vehicles is the industry's holy grail. "Look around," he said in late 2003, "just about everyone is doing something related to command-and-control. That aspect of the industry is very competitive right now. It's worth millions."[105]

Five Bombs in One Hole, and Cheaply
The Joint Direct Attack Munition and the Mass Production of Precision Destruction

Thus the military-industrial complex now consists of two relatively thin bookends to our enormous, civilian, high-tech economy. Military R&D programs push the leading-edge development of power semiconductors, software and sensors, a decade or so out ahead of Intel, Motorola or DaimlerChrysler, then encourage the migration of successful technologies out into the civilian sector as quickly as possible. Military contractors end up buying back the same technology at mass-production prices, embedding it in every vehicle, weapon and projectile on the battlefield.[1]

Introduction

On 26 October 2001, Master Sergeant John Bolduc, the officer in charge of Detachment Alpha 565 of the U.S. Army's Green Berets, presented his credentials to General Atiqallah Baryalai Khan, deputy defense minister of the Northern Alliance. Despite nineteen days of bombing by aircraft of the U.S. Air Force, the Royal Air Force, and the U.S. Navy, the war was still not going well for the enemies of the Taliban. General Khan was somewhat distressed by the small force he had been sent.

"Commander John," he asked, somewhat overstating Bolduc's rank, "there are only ten of you. When is the U.S. Army coming?"

Bolduc told him not to worry: "We're here. This is it."

Khan was not amused. "This is not good," he insisted. "We need many soldiers."

Bolduc was unfazed and reassuring. "You don't need many soldiers. Wait and see what we can do."[2]

Master Sergeant Bolduc was placing his faith in the ability of his team to serve as forward observers for a huge and concerted precision bombing campaign. Over the next two months, day after day, Detachment 565 spotted Taliban and al Qaeda positions, found their positions with GPS-laser rangefinders, and systematically passed those coordinates to strike aircraft overhead. Within minutes, 2,000-pound precision-guided bombs would typically fall within meters of the intended targets. At that distance, 2,000 pounds of high explosive is enough for any field fortification that their enemies could construct. Khan's troops were shortly exploiting large holes blown in the opposing lines.

From October through December 2001, Detachment 565 was the most active such unit in the country, calling in more air strikes than any other, and receiving credit for the deaths of 1,300 Taliban and al Qaeda fighters.[3] Dropping GPS-guided weapons out of intercontinental bombers for close air support was, in the estimation of Air Force Undersecretary Peter Teets, the single biggest contribution to the war effort, arguing that "the very idea that a B-52 or a B-1 flying around at 50,000 feet altitude could provide close air support to troops on the ground is a remarkable thought. Even as recently as ten years ago when we were involved in the Gulf War, such a thought would not have occurred."[4] What made this concept—indeed, the rapid victory over the Taliban—possible was a bomb: the Joint Direct Attack Munition, or JDAM (pronounced 'Jay-Dam').

The Requirement and the Technology

Despite the media attention of the last decade, precision-guided weapons are not a new concept. Interested in bombing German industry and railway junctions in the First World War, the U.S. Army Air Service attempted to build a gyroscopically-guided cruise missile out of a biplane. In the Second World War, the Luftwaffe and the Royal Air Force each developed practical radio-navigation systems for guiding bombers to their targets through darkness and adverse weather. The U.S. Army Air Forces also experimented with radio-controlled bombers that could be crashed, packed with explosives, into particularly well-defended targets. The Luftwaffe further developed a series of radio- and televisually controlled glide bombs for attacking ships: in addition to damaging two U.S. cruisers off Italy in 1943, the German air arm sank the Italian battleship *Roma* and a considerable tonnage of merchant shipping.

In the 1950s and 1960s, large sums were spent on inertial and astral guidance systems for nuclear weapons. Spending on precision guidance for more conventional weapons, however, was much more restrained in this period. The two types of precision-guided weapons available—laser-guided bombs (LGBs) and electro-optical guided bombs (EOGBs)—were also subject to significant operational and economic limitations:

Expense. While these guidance systems were reasonably accurate even by today's standards, they were also very expensive, so these technologies could only be applied to the most powerful weapons.

Vulnerability. LGBs and EOGBs required attacking aircrews to loiter within line of sight of their targets until the weapons had struck their targets. However, during the same time frame that these weapons were being developed, radar-guided surface-to-air missiles were replacing large-caliber anti-aircraft cannons as the primary means of high-altitude air defense around the world. This was forcing air forces to attack at lower altitudes—from which precision weapons could not be effectively employed, as they would not have sufficient time to steer themselves toward their targets in unpowered flight. Powered weapons—like cruise missiles—could be used for the most challenging targets, but these were generally too expensive for universal use.

Weather. That clear line of sight required clear weather: LGBs and EOGBs were difficult to use in fog, smoke, sandstorms, overcast skies, and even partly cloudy conditions. NATO countries were thus not likely to bet their defense against the Warsaw Pact on these types of weapons, since the weather in central Europe was frequently not conducive to visual bombing.

Toward the end of the Vietnam War, the USAF and the U.S. Navy had considerable success attacking bridges and vehicles with LGBs and EOGBs like the GBU-12 Paveway and the AGM-62 Walleye. After the war, however, frustration with the aforementioned limitations led most air arms around the world to limit their procurements of these two broad types. So, when U.S. aircraft attacked Syrian air defenses in Lebanon in 1983, or military and leadership targets in Libya in 1986, most carried unguided bombs. In the 1991 Persian Gulf War, only 7 percent of all U.S. air-delivered ordnance was precision-guided—even if intelligent targeting systems on the aircraft improved the accuracy considerably. The record of precision bombing in that war was impressive and impressively supported by video footage, but it was less impressive on cloudy days. The winter of 1990–91 saw the heaviest weather in the Middle East in eleven years, so Coalition bombers were often unable to find their targets.

After the war, USAF Chief of Staff General Merrill McPeak sent a hand-written memorandum to Major General Joe Ralston, USAF director of tactical programs, insisting that the service needed adverse weather precision munitions—smart bombs unaffected by clouds, rain, fog, or the thermal smoke of burning oil wells. Many in the USAF leadership at the time presumed that this meant a radar-guided bomb, but such a weapon could cost several hundred thousand dollars apiece. A team of engineers at the USAF Weapons Laboratory at Eglin Air Force Base, however, had another idea. In the mid-1980s, Lou Cerrato, a senior engineer there, had researched the concept of guiding bombs with inertial navigation systems (INSs). The idea had not attracted much funding: INSs drift with range, so only the most expensive ones were useful for long-ranged weapons.

For its part, the GPS program office had weapons guidance in mind right from the inception of the program. As Colonel (Dr.) Bradford Parkinson, the first USAF program manager for GPS, wrote to his staff far back in 1973, "the goals of this program are to build a cheap set that navigates ($10,000), and to drop five bombs in one hole." Weapons guidance had also been the primary motivation for the development of the U.S. Transit and Soviet Tsikada satellite navigation programs. Submarines carrying long-range missiles at sea could only attack their targets accurately if they knew their own positions with greater accuracy.[5] Transit and Tsikada, however, had obvious operational shortcomings: with only a handful of satellites in low orbit, signal availability was limited. GPS, with at least twenty-one satellites in high orbit, would guarantee almost constant access to satellite guidance.

The thought of depending for weapons guidance on signals from satellites 20,000 kilometers away failed for years to garner much enthusiasm. However, the example of the 1991 campaign in Iraq demonstrated what GPS could do, so General Ralston recommended adding a GPS receiver to the weapon,[6] and insisted that efforts to expand the program's very basic list of requirements (see below) be "stiff-armed." Cerrato's team tested a prototype in the winter of 1992, flattening an outhouse with an inert warhead from an F-16 at 20,000 feet.[7] The first JDAMs were delivered in 1997 for operational testing over the next two years. More than 450 were dropped in tests—the average miss distance was less than ten meters, and the weapon was reliable in 95 percent of the drops.[8]

INITIAL JOINT DIRECT ATTACK MUNITION (JDAM) REQUIREMENTS

Accuracy	When global positioning data are available, the kit must be able to direct the bomb to within 13 meters of the intended target.
Adverse weather capability	The munition must be able to achieve specified accuracy despite a range of inclement weather conditions.
Retargetability	The kit must permit in-flight retargeting of the munition right up to the moment when it is released from the aircraft.
Warhead compatibility	The kit must be compatible with the existing Mark-84 2,000-pound general-purpose bomb the BLU-109 2,000-pound penetrating bomb the Mark-83 1,000-pound general-purpose bomb and the BLU-110 1,000-pound bomb.
Aircraft compatibility	The munition must be compatible with the B-52H heavy bomber, the FA-18C/D Hornet multirole tactical aircraft, and the F-22 Raptor fighter (1,000-pound versions only for the latter two airframes).
Carrier operability	The munition must permit safe and reliable usage by aircraft operating from carriers at sea.

The JDAM at War

Integrating the JDAM onto a bomber or fighter-bomber is relatively uncomplicated, at least in comparison with the integration of other, more manpower-intensive weapons. The F-14 Tomcat provides a clear example: the U.S. Navy began to equip its F-14Bs to carry up to four JDAMs each in 2001, when the type had recently been armed with laser-guided bombs but was already going out of service. In 2002, the Navy modified its F-14Ds as well. The campaign against Iraq was their last major combat action: the Tomcats were even then being phased out of the fleet as new F-18F Super Hornet fighters rolled off a production line just across the Missouri River from the JDAM factory in Saint Charles, Missouri.[9] Modification made sense, however, because it was relatively easy and inexpensive: while preparing for action in Iraq in February 2003, a team aboard USS *Theodore Roosevelt* modified all the F-14s in the embarked air wing in seventeen days.[10] Since the weapon was so flexible, it was used widely in four wars that ensued shortly after its introduction.

KOSOVO, 1999

The B-2s of the 501st Bomb Wing kicked off the employment of JDAMs in 1999 by dropping 656 of the bombs in fifty sorties against Yugoslav infrastructural and command targets. The bomber crews flew missions lasting nearly thirty hours from their home at Whitman Air Force Base in Missouri. Since each bomber could carry up to sixteen 2,000-pound bombs on each missions, the twenty-one aircraft of the wing delivered 11 percent of the weight of all ordnance dropped on Yugoslavia during the war. Unfortunately, as further explained below, one of the strikes destroyed a wing of the Chinese Embassy in Belgrade, since the CIA targeting team that selected the building misidentified it as the headquarters of the Yugoslav federal arms import-export agency.[11] The mishap does illustrate the importance of accurate intelligence in the use of coordinate attack weapons. PGMs like the JDAM attack whatever is found at their target coordinates, regardless of whether it is the intended target or not.

AFGHANISTAN, 2001

The 2001 Afghan campaign, of course, was the war that made JDAM a widely recognized term, and in which victory was virtually assured by the JDAM. The USAF and the U.S. Navy used roughly 6,650 JDAMs in just two months of heavy combat,[12] a figure that accounts for over one-third of all the bombs dropped.[13] Roughly half these bombs were dropped by just eighteen USAF heavy bombers flying from the British base at Diego Garcia in the Indian Ocean. Assembly on the island was handled by just 154 munitions troops from the Mountain Home (Idaho), Ellsworth (South Dakota), and Barksdale (Louisiana) Air Force Bases; the weapons were shipped there in just two ammunition ships, the *Major Bernard F. Fisher* and the *Cornhusker State*.[14]

The force that carried those 3,000 or so JDAMs was composed of just ten B-52 Stratofortress and eight B-1 Lancer bombers of the USAF's 28th Air Expeditionary Wing. The B-52s typically carried twelve JDAMs on pylons under their wings; the B-1Bs carried as many as twenty-four in their bomb bays, and sometimes dropped twenty-four bombs on twenty-four separate aim points in a single mission. Each airplane averaged one mission every other day that was twelve to fifteen hours in length, and in the process, dropped over half the ordnance (by weight) that was delivered by air in the campaign.[15] B-2s flew only six missions, all in the first few nights of the war, all with JDAMs and all from their home base in Missouri. The targets were Taliban early-warning radars and military headquarters.[16] The withdrawal from the campaign of the B-2s was unsurprising. After the first night, there were essentially no Taliban high-altitude air defenses left, so B-52s and B-1s could range over the country unmolested. Moreover, overseas basing for the B-2s has been difficult, since the bombers require extensive maintenance to their stealthy coatings after every flight, and the trip from Missouri to Diego Garcia is thirty hours in each direction.[17]

This massive use of the JDAM completely altered the nature of the war. As noted above, this concentrated firepower was used to blow holes in Taliban trenchlines for the troops of the Northern Alliance to exploit. Major General Dan Leaf, who after the war was director of operations requirements for the USAF, suggested that the psychological shock of the bombardment upon the troops of the Taliban and al Qaeda induced a rout once they realized that the capabilities of U.S. and British airpower were overwhelming.[18] Some of those on the ground put it more emphatically. After four JDAMs put the main runway at Kabul International Airport out of action, Farid Ahmad, the chief of the crew attempting to repair the damage, exclaimed, "I have been through the Russians. I have seen [warlord Gulbuddin] Hekmatyar in action and the Northern Alliance. This is just incredible. The Americans appear to have been 98 per cent accurate."[19]

The particular advantage of the intercontinental bombers was not just their range but also their staying power. Measuring their loiter time in hours rather than minutes (like fighter-bombers), the B-52s and B-1s could be counted available almost whenever they were needed along the northern front.[20] The bombers, however, were not enough—it was the presence of U.S. and British forward observers on the ground with the troops of the Northern Alliance that guaranteed that the strikes would almost always be accurate—75 percent of the time, as the U.S. Navy initially assessed. This was much better than the rather low rates achieved in Kosovo, where the lack of ground observers seriously hampered targeting.[21] In Afghanistan, given the short preparation for the campaign, this initially did not go so well. Army Green Berets had not had significant experience calling in air strikes and actually deployed to Afghanistan without USAF combat controllers (dedicated forward observers). Few had given much thought to using JDAMs for close air support. Under the strain of combat, that changed quickly: the USAF commander in the Middle East lent the Army the same teams that would have supported line regiments in combat, and the JDAM quickly proved very useful in this role.[22] After the war, to facilitate targeting, Special Operation Command ordered the development of the Joint Expeditionary Digital Information system (JEDI), a Windows CE handheld computer that takes data from laser-rangefinding binoculars and allows a trooper to input additional data (such as target type) by touching a series of screen icons. The machine then communicates with strike aircraft by sending text messages via an Iridium telecommunications satellite.[23]

When a wider envelope of destruction was required, an INS-guided Wind Corrected Munitions Dispenser (WCMD) could rain thousands of cluster bombs on massed enemy troops. When the targets were too small or moving too quickly, a laser-guided bomb would do the job.[24] The largest part of the destruction, however, was accomplished with JDAMs. Since the JDAM simply flies to a point in space, almost regardless of weather, the spotting team requires merely a single observation of a stationary target. Once the geographic coordinates have been passed to an attacking aircraft, the team can turn its attention to identifying and geo-locating its next target. The ease of this process in combat drastically increased the number of individual bombs that could be successfully dropped and the number of Taliban and al Qaeda troops that could be killed on a given day.

Success with the JDAM—and an understanding of its further potential— put the weapon in economic demand. Within a week of the initial attacks on the United States in September 2001, the Department of Defense sent an emergency war appropriation to the Congress with nine items. One, for $4.6 billion, was for more additional JDAMs; that sum was enough to procure over 200,000 guidance kits. The Pentagon also asked Boeing and its subcontractors to immediately ramp up production of the weapons—without waiting for formal Congressional approval of the appropriation. The paperwork would follow, but the bombs were needed immediately.[25]

IRAQ, 2003

By the time of the war in Iraq in 2003, U.S. forces had roughly 20,000 JDAMs in inventory.[26] This number proved more than adequate—with so many on hand, the JDAM was described as the "weapon of choice" by Colonel James Kowalski, commander of the B-1 bombers of the USAF's 405th Air Expeditionary Wing. He specifically noted that "when you mate it up with a B-1 that can carry twenty-four of them and then basically range across Iraq, you can hold at risk just about any target in the country."[27] Accordingly, JDAMs accounted for about 29 percent of all air-delivered ordnance dropped during the campaign.[28] Laser-guided bombs (particularly the Raytheon Paveway series) were used somewhat more heavily,[29] but this was because, as Major General Dan Leaf (who was by then the USAF liaison to Lieutenant General William Wallace, Coalition Land Forces Component Commander) put it, "we have them in the inventory in great numbers . . . the terminal guidance for refining targeting in a visual engagement is very good," and "it is very good in close contact." He also extolled the importance of the JDAM's operational simplicity, noting that "in the fight near An Najaf, with elements of the 3rd Infantry Division in the dust storm at night, having bombers with dynamically retargetable precision-guided [JDAMs] in the 2,000-pound class was exactly right. They were persistent. They were mass-targetable and against massed enemy troops and armour and vehicles they were very effective."[30]

At this point, the B-1s were conducting a two-front war, bombing targets in Iraq and taking calls for fire against insurgents in Afghanistan.[31] On 7 April 2003, one B-1 crew dropped four JDAMs in a well-known attempt to kill Saddam Hussein and his sons as they supposedly met in a Baghdad restaurant. The effort failed, as the intelligence was faulty, but the bomber completed the strike only twelve minutes after receiving its target coordinates, which indicates the great flexibility of the weapon. The crew later dropped twenty more JDAMs on seventeen additional targets north of Baghdad to complete its ten-and-one-half-hour sortie.[32] F-16s and F-18s also used JDAMs, in addition to a variety of other weapons such as CBU-87 and CBU-103 WCMDs, AGM-65 Mavericks, and AGM-88 HARMs.[33] Royal Air Force Tornado GR4s did not have JDAMs, but instead used a mixture of 265 laser-guided Paveway and 404 GPS/INS laser-guided EGBU-27 Enhanced Paveway IV guided bombs in Iraq.[34] With all these coordinate-attack weapons flying about, many of Iraqi soldiers decided not to wait to discover what might happen. In one combat engineering squadron of the Republican Guards' Hammurabi Division, 145 of 150 troops deserted before U.S. troops had arrived near their positions.[35]

The RAF's decision to buy a combination laser-INS/GPS-guided weapon was followed by a similar commitment from the USAF. For the war in Iraq, five hundred GBU-27s were upgraded for the roughly fifty-plane F-117 Nighthawk wing since JDAM integration on the light stealth bomber was not scheduled until fiscal year 2006.[36] In addition to dropping JDAMs, B-2s often flew against Iraqi targets with 5,000-pound GBU-28 bunker busters, initially developed for the 1991 campaign, but which by 2003 were also GPS/INS-guided.[37] The USAF's other huge bomb was the MOAB—Massive Ordnance Air Blast, or (more popularly) the Mother of All Bombs. Dropped from a C-130 Hercules, at 21,500 pounds, this was the largest non-nuclear warhead available, with 50 percent more explosive power than the Daisy Cutter that was used in both Vietnam and the 1991 campaign.[38] The MOAB, too, was GPS-guided, even if the huge warhead rendered that extreme accuracy less important.

LEBANON, 2006

By 2004, a wide selection of air arms was procuring JDAMs, from the USAF[39] and the U.S. Navy, through the Royal Australian Air Force,[40] to the Israeli Air Force. The Israelis ordered 5,000 JDAMs from Boeing in the summer of 2004, including 500 specifically made for the 2,000-pound BLU-109 bunker-busting bomb, for a total of $319 million. These came in handy in the summer 2006 during the war with the Hezbollah Islamist insurgent group. After spending two years planning an expansive bombing campaign, using both human intelligence and imagery from the Ofek-5 reconnaissance satellite, the Israeli Air Force made wide use of the JDAM against Hezbollah fighting positions in southern Lebanon and headquarters facilities in Beirut.[41]

The Business of the JDAM

The JDAM program achieved this ubiquity through not just its flexibility but also its low price. In 1992, USAF Chief of Staff General Merrill McPeak told Terry Little, the first JDAM program manager, that he only wanted the weapons if they could be bought for less than $40,000 apiece.[42] Initial estimates based on historical experience had suggested that the price would be $68,000, but Little knew from long experience as a government weapons procurement manager that he could do much better. Lower prices could be achieved by eliminating many government business practices that added costs to contractors' efforts.[43] As Little later told a chronicler of the JDAM program, not all the problems were statutory—many were simply the result of excessive bureaucracy in governmental decision-making. "In 1993," he said, "to get the project started, I gave forty-eight briefings to senior people who were not in my chain of command. Our program-approval documentation was literally six-feet high and took 10,000 man-hours to prepare. And this was for a program that was not a technological challenge, was a high priority, and was uncontested. This was business as usual at the Department of Defense."[44]

Simplifications of regulatory requirements were permitted by the Federal Acquisition Streamlining Act of 1994, which authorized the secretary of defense to waive any regulation not required by statute that might impede the efficiency of the contracting process. Over the course of the JDAM's development, the program office was exempted from twenty-five provisions of the Federal Acquisition Regulations and a further twenty-five provisions in its Department of Defense supplement. The hallmarks of this approach can be summarized in nine points:[45]

> *Rolling down-selection.* In April 1994 the USAF and the Navy selected Martin Marietta and McDonnell Douglas from among seven bidders to build prototype guidance kits. Over the following eighteen months, the two companies competed in the first phase of engineering and manufacturing development effort. The two-phased competition, common in complex commercial procurements, allowed the government to concentrate its managerial attention on the two most competitive teams.

> *Primary award criteria based on past performance and best overall value.* Previously, military procurement authorities in the United States had often been prevented from considering a company's track record in evaluating its ability to produce a new weapon. Further, selections had to be made on very technical criteria that theoretically minimized the role of the government's managerial experience and intuition. Unfortunately, this encouraged many upstart organizations to file proposals with the government that required costly evaluations by large program management staffs. If the JDAM program office was to be limited in staff (to minimize its costs), a more sensible approach commonly followed in commercial procurement would be required.

> *Government/supplier integrated product teams (IPTs).* After that initial down-selection, the design teams at both contractors were assigned engineers, contract managers, and program analysts from the Air Force and Navy departments to assist them in their bids. Proverbial Chinese walls were erected between the teams to prevent them from leaking information to one another—in this way, the military staffs were encouraged to do everything necessary to help their respective contractors win the competition by showing them firsthand what the government needed in the weapon and its supporting systems. The objective was to make each team as competitive as possible so that the military services could procure the most cost-effective weapon possible.

Performance-based, head-to-head competition. Indeed, the IPTs held a critical role in ensuring the best product for the forces. McDonnell Douglas's team initially presented a plan for a weapon that would have cost about $28,000 in nominal 1995 terms—roughly <$10,000 more per unit than what Lockheed Martin was proposing. After a warning from the USAF lieutenant colonel who was leading their government assistance team, McDonnell's people set to cut costs drastically by combining the functions of multiple parts. In the end, the McDonnell team got the cost down to $14,000 per unit, or half what it had initially proposed.[46]

Contractor-supplied warranty. That $14,000 cost (in then-year terms) included something that almost no other weapons had before—a warranty. At the end of the production line, JDAM tail kits are packed in foam, sealed inside vapor-sealed bags, boxed in twos, and given a twenty-year performance guarantee.[47] This contract provision considerably simplifies the government's administrative burden, lowering its total cost of ownership by allowing munitions troops to concentrate on operational and logistical matters rather than issues related to materiel management. Arguably, the contractor who built the weapons possesses information about the serviceability of the systems that is at least as good as that had by the government, so the arrangement is economically efficient.

Minimal paperwork and limited, streamlined oversight. Waiving cumbersome procurement rules brought low-cost, commercially oriented firms into the bidding for JDAM parts. The strakes and cable covers are made by Stremel Manufacturing of Minneapolis, which had no military contracts before the JDAM, instead making components such as lawn mower bodies for Toro.[48] (A partial list of major suppliers follows in fig. 4.6.)

Use of commercial products. Even though the tail kit was a military article, 85 percent of its value was in components that had commercial analogs, including its Honeywell inertial measurement unit, its Rockwell Collins GPS receiver, its Lockheed Martin mission computer, and its Textron tailfin actuators.[49] Many of the other components—wings, wiring harnesses, and metal structures—were the products of well-understood production processes. Treating the weapon in a quasi-commercial manner allowed the program office and Boeing very quickly to tap into technological advances. For the JDAM, the most significant element in the cost reduction of the weapon over the course of the program was the drop in the price and weight of the GPS receiver.[50] This was a by-product of commercial investment in GPS technology *outside* military programs, as Peter Huber noted in *Forbes* (see above). Too closely specifying the nature of the guidance system in advance would have shut off the benefits of these investments from the government.

Allowing trade-offs of price for performance criteria. Except for a few live-or-die criteria (outlined above), the program manager had wide authority to procure the most cost-effective weapon possible. In theory, the U.S. military services had often established formal, written requirements for weapons without considering the costs of meeting the individual strictures of those statements. *Ex post*, program analysts would modify programs in the breech according to cost-effectiveness, but this process was far from administratively efficient.

Firm, fixed-price production contract. Once the cost target had been set, the prospective contractors worked quite hard to secure the business that this promise represented. As Kim Michael, Boeing's (former) program manager, put it, "we were willing to take a risk up front to get a long-term commitment to a price."[51] While the JDAM program would produce a vast number of weapons, it would not produce so many as to justify maintaining two production lines. Estimates of the microeconomics of production indicated that the overall costs would be minimized with a single line, even in the absence of follow-on competition. This provided an enormous incentive to the competing contractors to get the pricing right from the start, which would also provide programmatic stability.

MAJOR JOINT DIRECT ATTACK MUNITION (JDAM) SUPPLIERS

Part	Supplier	Location
Tail actuator	HB Textron	California
Power supply	Lambda Advanced Analog	California
GPS antenna	Aero Antenna	California
GPS antenna cable	C.E. Precision	California
Guidance unit chassis and cover	Hyatt Die Cast	California
Container	Plastics Research Corporation	California
Tail control fin	Precise Machining, Alcoa Forged Products	Oklahoma
Strakes and cable covers	Stremel Manufacturing	Minnesota
Hardback	Progress Casting	Minnesota
Intertial measurement unit	Honeywell	Minnesota
GPS receiver	Rockwell Collins	Iowa
Power supply	Modular Devices	New York
Tail fairing	Lockley Manufacturing	Virginia
Wiring harness	Woven Electronics	South Carolina
Battery	Enser Corporation	Florida
Mission computer	Lockheed Martin	Florida

Note: From Ingols and Brem, in *Implementing Acquisition Reform,* exhibit 17. Available at http://www.arnet.gov/comp/seven_steps/library/JDAMsuccess.pdf.

Even with all these measures in place, two other departures from standard practice were essential to this program. Without them, much of the rest of the reforms would make little sense.[52] The first was contractor control over the technical data. Prior to the JDAM program, most weapons procurement contracts in the United States stipulated that the government would own most of the significant intellectual property that resulted from the development of the system. Control of the technical data was intended to guarantee that the government could not have its national security interests held hostage by self-interested contractors, and that the government could use those data to shop for more cost-effective contractors in later years. The problem was that the loss of control also pulled most of the incentive for performance improvement from incumbent contractors, because any new ideas that they devised could easily be stolen away by the government and given to competitors. With the JDAM, if the contractor was to be motivated to provide the lowest price possible up front, then the program office would have to guarantee that its achievements would remain in-house. The second difference involved negotiations based on supplier price, not cost. Further, the government would have to agree that the contractor's price was its final offer. This is not to say that the USAF would not be interested in understanding the competing contractors' negotiating positions, but rather, that it would base any final contract agreement on an absolute number. Many weapons procurement contracts in the United States, particularly developmental ones, had stipulated that the government would pay the contractor's cost (determined in many cases by large and intrusive audit staffs) plus a profit margin to compensate the contractor for its cost of capital. This also restrained the contractor's enthusiasm for cost-performance—reducing costs meant reducing absolute profits, and there has never been much economic point to that.

In October 1995, McDonnell Douglas was selected to complete the development of the weapon and its production system, substantially on its low bid and its guarantees of progressive, inflation-adjusted price reductions over the course of the contract.[53] Despite all this effort, however, the JDAM was only just ready for its role in Kosovo, and mandated governmental bureaucracy was the culprit. The authority of the SPO to waive the statutory requirements had a sunset clause—which expired while the weapon was on the verge of low-rate initial production, and just as the build-up for the campaign was beginning. With predictions of foul weather and a prohibition on the use of ground forces restraining the expected efficacy of laser-guided bombs, the SPO "had to bend every rule in the book" at the specific direction of Joint Chiefs of Staff. Influential staff on Capitol Hill agreed to look the other way for the duration of the war,[54] and the Congress wisely extended the authority after the war. This mandated lack of trust (more the fault of the Congress than the Pentagon) in military contractors thus has a very real cost.

Building the Bombs: The JDAM Factory and Supply Chain

By the time that full rate production was approved in March 2001, after Boeing had demonstrated that the production line was flowing very efficiently, the price per bomb kit stood at $21,000 (in inflation-adjusted terms).[55] Attaining this impressively low price required concerted managerial attention amidst expanding production requirements brought on by significant regional wars in Afghanistan and Iraq. As noted above, within a week of the September 2001 attacks on the World Trade Center and the Pentagon, the JDAM program office asked Boeing to begin increasing its JDAM production rate and to worry about the contractual details later.[56] In the summer of 2002, Boeing built a 30,000-square-foot addition to its factory in Saint Charles, Missouri (an outer suburb of Saint Louis), to handle the expanded demand.[57] By the summer of 2003, Boeing was producing 3,000 JDAMs per month at its factory in Missouri and had produced a total of more than 60,000 bombs.[58]

Boeing's supply chain operations are quite streamlined. At the plant, parts from the twenty-two primary component suppliers arrive continuously on trucks that Boeing controls and tracks continuously with GPS receivers. To minimize inventory carrying costs, the factory keeps no more than six days' supply of parts on hand. To minimize handling costs, all the parts arrive in reusable containers, and the physical flow of inventory is last-in, first out (LIFO)—most incoming parts move immediately to the production line.[59] From here, about forty production-line workers assemble the kits. All have a common job classification: munitions mechanic. Despite a long history at Boeing of labor disruptions due to disagreements over work rules,[60] the machinists' union at Boeing agreed to this cost-saving simplification to help secure the contract. Another 150 staff work off the production line in administrative and logistical capacities. On the line, parts are combined at fifteen stations to produce a tail kit in about ninety minutes.[61]

This is not to say that the supply operation has been trouble-free. In March 2003, Nicholas Hayek, president and majority owner of well-known watchmaker Swatch, told Swatch subsidiary Micro Crystal to halt shipments of oscillators needed in GPS receivers to Honeywell, a subcontractor on the program to Boeing. Hayek was somewhat miffed over the U.S. invasion of Iraq. Boeing called the Pentagon, the Pentagon called the White House, the White House called the Swiss federal government, and the Swiss government instructed Swatch to honor its contractual commitment. The oscillators were dual-use products, not exclusively military ones. Under Swiss law, shipments of dual-use products to belligerents is legal in wartime, so the contracts had to be honored. In the interim, however, Boeing needed to buy oscillators from a U.S. firm at nearly twice the price.[62] The reaction from U.S. Congressman Duncan Hunter was quite sharp, suggesting in effect that firms outside the United States be removed from the supply of critical components for U.S. weapons.

Fortunately, cooler heads have since prevailed. The response from the JDAM program office, from the Pentagon's industrial policy office,[63] and the Bush Administration as a whole was more measured, as at least one overseas supplier—another microchip manufacturer—had performed yeoman service for the program in a past war. In readying JDAMs for the Kosovo campaign, regulations and statutes were not the only hurdles that the USAF faced—some of the problems were physical. In early 1999, Boeing was asked to assess its ability to rapidly expand its trickle of production to support the gathering war effort, taking into consideration any limitation attributable to supply constraints. The limiting factor was determined to be the supply of tuning crystals for the GPS receivers. The best source was Navman, a small, private maker of navigation equipment in Auckland, New Zealand, that had built its own microchip fabrication plant a few years before in Christchurch. The company had developed a proprietary process for rapidly aging new crystals to achieve the desired level of timation stability. Some in the USAF were understandably nervous about buying an essential part overseas from a firm with no track record as a U.S. military supplier, but since Navman's process ran an order of magnitude faster than anyone else's, there was little reason to get another firm involved. The company delivered enough crystals to build enough bombs to keep the USAF's B-2 wing supplied for the war, and was well paid for its exertions. Indeed, Navman's efforts attracted considerable attention. Immediately after the war, Darlene Druyun, then the USAF's chief weapons buyer, traveled to New Zealand to present the owners with an award (and Druyun is known to hate business travel).[64] In June 2003, Brunswick Marine, a large Wisconsin-based builder of boats and marine electronics, bought a controlling interest in the firm acquired the rest by the end of 2005, and then sold the operation to MiTAC of Taiwan in 2007.

The Limitations of the JDAM

Despite enormous enthusiasm for the JDAM, it must be admitted that coordinate-attack weapons are not fully suitable for evasive, mobile targets. This is not to say, however, that the problem is intractable. The USAF's Affordable Moving Surface Target Engagement (AMSTE) program is seeking to use multiple radars to cue JDAMs against moving, though not necessarily maneuvering, ground targets. JDAMs about to drop could be passed the expected target coordinates of their prey at impact based on the vector calculations of the targeting system.[65]

That said, producing target coordinates is not always a simple matter. When hostile troops are identified in the lines directly opposing one's position, a JDAM can often make short work of them. The problem is that not all targets are so readily identified, and misidentifications have led to some costly mistakes with JDAMs. As noted above, during the 1999 Kosovo War, a B-2 bomber dropped five JDAMs on a building that CIA targeting analysts had identified from satellite photographs and street maps as the Yugoslav Federal Directorate of (military) Supply and Procurement. The 2,000-pound bombs wrecked an entire wing of the building and killed three people inside. The problem was that the building in question was actually the Chinese embassy. A series of errors had led analysts to misinterpret the imagery. This indicates the great but self-limiting advantage of coordinate attack systems: their ability to seek out targets based solely on their locations. While this keeps friendly forces high above ground fire, it also creates an enormous intelligence collection and analysis burden associated with producing the extremely accurate targeting intelligence that these weapons require.

Fratricide and collateral damage are also lingering problems for most precision weapons, and this includes JDAMs. Three incidents in the 2001 Afghan campaign illustrate some of the problems: On 12 October 2001, a JDAM dropped by an F-18C against a Taliban helicopter on an airfield near Kabul instead hit a house in a neighborhood about a mile away.[66] Four were killed on the ground and eight were wounded. The miss distance was almost certainly too large for a system malfunction (these have been known on rare occasions to happen). Rather, the error was human. The laser spotting system and the JDAM targeting system in the F-18C had not been fully integrated at the time, so the pilot had to type the target coordinates that he saw on the laser-spotting system's readout back into the targeting computer that controlled the JDAM. In typing, he transposed two digits, and sent the bomb onto another small-area map in the U.S. military's Universal Transverse Mercator mapping system. Fortunately, this problem was later rectified with the Hornet Autonomous Real-time Targeting (HART) upgrade for the F-18E/F Super Hornet. On 25 November, several friendly Afghan soldiers were killed and five U.S. troops wounded by a JDAM that landed too close to them during the prison rebellion at Mazar-i-Sharif. Artillery in close contact with friendly forces often leads to fratricide, but 2,000-pound bombs can cause even bigger problems if they drift a bit too close. On 5 December, a JDAM dropped by a B-52 killed three Green Berets and wounded nineteen other U.S. and Afghan troopers.[67] In this case, the GPS receiver that the USAF combat controller on the ground was carrying experienced a battery failure just as he was readying the target coordinates for his call to the bomber overhead. The controller replaced the batteries, turned the receiver back on, and notified the bombardier of the coordinates. The problem in this case was that the receiver had reset itself after the power failure, so the coordinates on the readout were his own. Dutifully, the air crew dropped a JDAM virtually on top of him.

The low cost of the JDAM's GPS/INS system also suggests a potential vulnerability to electronic jamming. Jammers capable of interfering with GPS are rather easy and inexpensive to construct, at least by the standards of military electronics.[68] In May 2003, Iraqi forces briefly attempted to jam GPS reception in metropolitan Baghdad, presumably to interfere with strikes using JDAMs and the longer-ranged Joint Stand-Off Weapon (JSOW). Coalition forces quickly identified seven jammers and destroyed them with air strikes—one, according to Air Force Secretary Jim Roche, using a GPS-guided weapon. Nonetheless, the U.S. military has been taking the GPS jamming problem quite seriously for over ten years, and in January 2003, awarded a $50 million, three-year contract to add anti-spoofing and anti-jamming capabilities to the GPS receivers embedded in JDAM kits.[69]

The JDAM and Transformation

Proofing the JDAM to some extent against this sort of attack could be important because the JDAM has such a central role today in the United States' (and others') military capabilities. Immediately after the invasion of Iraq, the Pentagon had huge quantities of the weapons on order: 100,000 2,000-pounders, 90,000 1,000-pounders, and 40,000 500-pounders.[70] The 500-pounders, originally of interest mostly to the USMC, more recently became of greater interest to the USAF. In March 2003, during the advance of Coalition troops into Iraq, the USAF and Boeing flight-tested a B-2 with 500-pound GBU-38 JDAMs at Edwards Air Force Base.[71] This is significant because the 500-pound warhead is particularly valued in close air support of ground troops, and because the B-2 could carry eighty of the weapons at a time. Indeed, Northrop Grumman and the USAF tested this concept in September 2003—a B-2 crew dropped eighty 500-pound JDAMs against eighty targets in a single twenty-two-second pass on Hill Air Force Base's Utah Test and Training Range.[72] This is simply a shocking quantity of firepower to put on a single aircraft.

This potential for the JDAM as a flying artillery weapon considerably explains on its own why former Defense Secretary Donald Rumsfeld elected to cancel the Crusader howitzer project, as well as why U.S. Army Chief of Staff

General Peter Schoomaker decided to cancel the Comanche attack helicopter project. Still, even 500-pound warheads are overkill for many field targets, given the accuracy of the JDAM. If targets could be killed with smaller bombs, then aircraft could carry more weapons, destroy more targets, and fly further on the same fuel loads as before. According to a 2002 study by RAND, the single most cost-effective thing that the Pentagon could do to improve the fighting capacity of U.S. forces would be to introduce a small (250-pound) GPS/INS-guided precision bomb. Indeed, it was estimated to exceed the cost effectiveness of the next most marginally cost-effective option by a factor of *four*.[73] Accordingly, shortly after the Afghan campaign began to wind down, the USAF held a competition for the procurement of some 24,000 250-pound Small Diameter Bombs. After a two-year competition, Boeing beat its recurring competitor Lockheed Martin to secure the first phase of a program that will eventually lead to roughly $2 billion in business for one or more bomb makers.[74] The weapon first flew in a flight test in February 2003,[75] and its GPS guidance system is expected to be considerably more jamming-resistant than that on the JDAM. By 2005, Boeing was producing the weapons at the same factory that produces the JDAM, as the design of the weapons are just too similar to avoid the synergies of co-location. The program, expectedly, features some of the same suppliers as with the JDAM, including Textron, Honeywell, and Rockwell Collins.[76] The program is expected to run to approximately 24,000 weapons.[77]

The Future of Weapons like the JDAM

Weapons like the JDAM have considerable potential to help close the transatlantic gap in military capabilities. In April 2002, Chairman of the U.S. Joint Chiefs of Staff General Richard Meyers wrote to a largely British audience that "the money saved by purchasing one or two fewer Eurofighters . . . could be used to acquire several thousand PGMs, resulting in a dramatic increase in combat capability."[78] The same month, the USAF announced that it had increased its expected total purchase of JDAMs to 236,000 weapons kits.[79] If European air forces had one-tenth that many precision weapons, their contribution to allied military operations would be drastically increased.

Unfortunately, up through the time of the Kosovo campaign, PGM purchases by the European members of NATO did not reflect the great importance of this class of weapon in modern combat. PGMs may have constituted only about 9 percent of air-launched weapons in the 1991 campaign in Iraq, but the use of laser-guided bombs to destroy Iraqi tanks and artillery was very significant.[80] In the 1995 Bosnia, 2001 Afghan, and 2003 Iraq campaigns, PGM usage averaged about 70 percent of all air-dropped weapons usage. The 1999

Kosovo campaign featured only about 35 percent usage, as bad weather and limited PGM availability over an unexpectedly long campaign led a number of NATO air arms to use cluster bombs for attacking suspected Yugoslav troops.[81] In open country, the European air forces might fare just fine—there would be little compunction against using their rather plentiful cluster bombs—against a force of perhaps six mechanized divisions. For reference, Iraq invaded Kuwait in 1990 with eleven divisions. Anything more would a considerably larger investment in precision munitions, but this investment has not been forthcoming.[82] Between the 1991 Gulf War and the 1999 Kosovo campaign, European air forces ordered only about 3,700 PGMs.[83]

This is not to say that the U.S. air arms could not improve as well. The USAF recently began development of the JDAM-ER (extended range)—a bomb with a range of 80 km, approximately triple its current performance, with the same accuracy. The per-unit cost target of the wing kit is $10,000. A flight test in April 2000 using MBDA's Diamondback wing kit on a GBU-31 JDAM validated the concept.[84] The U.S. Navy and USAF together have an ongoing interest in equipping some JDAMs with terminal seekers. This would not necessarily defeat the purpose of a coordinate attack weapon—rather, the highly accurate GPS/INS navigation should get the seeker to the point that a relatively inexpensive one can get the job done. Accordingly, the Navy's JDAM product improvement program (PIP) would add a commercially derived, uncooled, imaging infrared seeker to the nose of the JDAM that would activate only a mile from the target.[85] Extensions to the JDAM program, however, further illustrate how it has come to dominate PGM planning in the U.S. military.

Large Quantities, Small Factory, Big Company: The JDAM as a Case Study

Since Boeing may have locked up a considerable portion of the market for low-cost PGMs, it is worth questioning how the company secured and maintained that position. Despite the radical nature of the concept, and its almost radical simplicity, the JDAM is not, by the criteria laid out in chapter 3, quite a program in which a small or medium-sized firm would have a relative advantage:

INNOVATION WITHOUT R&D INTENSITY

The JDAM is without question a radical innovation in the field of precision weapons, but it was not an R&D challenge. As Terry Little noted, the weapon was far from a technical stretch once GPS had been shown to be reliable in combat. Integrating the parts was a matter of engineering, not fundamental research. However, there were problems with each of these three conditions that pertain for any small firms interested in entering the market.

First, the government had already innovated in the ways that were necessary in its tests at Eglin. Once Cerrato's team had demonstrated the concept, the basic outlines of the design were established: the JDAM would carry a multichannel GPS receiver, an inexpensive omnidirectional antenna, an INS incorporating a ring laser gyroscope, and a set of electrically actuated tailfins. Subsequent development work by the contractors was a matter of design-for-manufacturability. This is a critical competence for any manufacturer aiming to produce relatively commoditized products by the thousands to the hundreds of thousands. It is also one that is found in large aircraft and weapons manufacturers.

Second, while the challenge of the JDAM was not one of technology, it was one of marketing: the USAF and Navy leadership had to be convinced to allow satellite signals emanating from 20,000 kilometers away to guide their precision weapons, even in the face of electronic jamming with thousands of times their signal strength. Early on, the survivability and cost estimates for the JDAM faced considerable doubts in the national security establishment. Convincing skeptical opinion leaders required a concerted marketing and lobbying effort by people inside and outside the government, and established weapons makers (such as Boeing) hold advantages in these areas.

Finally, even though the innovation was relatively radical in its effects, it was hardly at all operationally. The JDAM would be dropped by the same aircraft that dropped laser-guided bombs—it would just be easier to employ. Once the weapon had been tested in combat, questions about technological trajectories considerably settled down. By the late 1990s, nearly every new weapon systems proposal carried the assumption of GPS/INS guidance.

SKILL-INTENSIVE PRODUCTION

Production on the JDAM line in Saint Charles certainly requires skilled labor, but not to the extent required for shipfitting or satellite construction. The common coding of all the machinists as weapons builders testifies to the efficiency of the line, but it is also demonstrates that the modular nature of the weapon makes its assembly more akin to automotive work than aerospace work. It is important to note in this case that there are rather few small automotive firms. Scale, in this situation, is a more important economic attribute than skill, even though skill is very important. So, while a considerably smaller firm could have established and managed a production line the size of that in

Saint Charles, a smaller firm would not necessarily have held a competitive advantage in doing so.

Boeing's contractually guaranteed price reductions testify to the speed of its learning curve and its confidence *ex ante* in its ability to attain that curve. Companies that are effective at building highly engineered, specialized electronics products are often not those that excel at cost-effectively building mass-market electronics products with steep learning curves. More so, the former route is not strictly the more profitable in an industry that features rapid technological change.[86] Computer hard drive manufacturers are constantly expanding the capacities of their products by large factors, but competition among them is so strong that few earn more than their cost of capital. To quote a Boeing manager working on a rather different program, the ideal aerospace product "has to be small, accurate, long-range, fast, and cheap. These attributes don't often show up at the same party."[87] The JDAM got everything but a long range, though altitude is protection in the absence of sophisticated air defenses. Boeing, however, particularly after its merger with McDonnell Douglas, had the skills both to build highly engineered products and mass-market ones. Those are the capabilities that are not often seen in one firm, and rarely at all in small ones.

In short, the JDAM concept was insufficiently innovative, too recombinative, and too cost-controlled over the long run for a small firm to have been very competitive. That said, the program demonstrates an important industrial policy lesson. If the government wishes to encourage small and medium-sized firms to attempt to garner prime contracts, then contractors must be allowed to set prices that do not require audits of their costs and to own the rights to the designs they develop. As noted, negotiating on cost limits a contractor's potential profits, which is superficially attractive to the public-minded, but doing so also limits the attractiveness of the market for the weapons in question. Demanding control of the technical data has a similar effect, in that weapons developers become effectively time and materials contractors with large capital requirements.

Small firms have difficulty competing in this environment. The small-cap phenomenon in stock prices explains why—small firms trade at a considerable discount to the prices that the capital asset pricing model (CAPM) predicts that they should fetch. This is because investors demand much more from assets with very high market-adjusted risk than from relatively stable issues. In the long run, this accounts for the high average returns attainable in technologically advancing markets supported by venture capital. Venture capitalists—or any others interested in making money—have little reason today to invest in small firms with intellectual assets with great military potential. If the government intends to punitively cap their profits and seize control of their ideas, there is insufficient upside to warrant an investment in a risky firm with a great idea. Since arms contractors and their investors do not make such ventures out of pure patriotism, the Pentagon and other defense ministries might seriously consider the long-term attractiveness of contracting regimes like the JDAM program.

5

Dili and the Pirates
HMAS *Jervis Bay* and the Military Potential of Aluminum Catamarans

> The U.S. military was, quite frankly, stunned by what that vessel could achieve. Personnel from the Seventh Fleet who encountered the vessel during East Timor peacekeeping operations had simply never seen the like of it.[1]

In 1520, Portuguese traders searching for sandalwood trees found the island of Timor and claimed it for King Manuel I in Lisbon. Portugal's attention to the island was slight over the next several centuries, and its sovereignty over the eastern half was confirmed only in the 1859 Treaty of Lisbon with the Netherlands (which controlled West Timor and the rest of the East Indies). In 1974, after a long period of administrative neglect, a rising separatist insurgency movement, and the fall of Marcelo Caetano's dictatorship, the Portuguese government abandoned the colony, recognizing it as an independent state. Rivalry between opposing political parties, however, soon led to civil war. In September 1975, Indonesian troops began moving in, and the Indonesian government under General Suharto annexed the territory as its twenty-seventh province in 1976.[2] The combination was not a harmonious one: while Indonesia was a large, Muslim, polyglot country under military dictatorship, East Timor had less than one million people, was 92 percent Roman Catholic, and had little connection to the military authorities in Jakarta 1,300 miles away. Unsurprisingly, guerrilla resistance to Indonesian rule continued for years.

After Suharto's dictatorship fell in 1998, his successor, President B. J. Habibie, offered the Timorese a choice between autonomy and independence. On 30 August 1999, 80 percent of voters chose the latter. Shortly thereafter, angry Indonesian soldiers and freelance unionists went on the rampage, sacking what they could. At sea, this could be called piracy; on land, it was mere mayhem. On 12 September, Australian troops began arriving in Dili, the capital, to reestablish order. The first ship to moor alongside the decrepit (and now damaged) port facilities was an aluminum-hulled catamaran car ferry that had been hurriedly chartered from the Tasmanian shipbuilder Incat (International Catamarans) by the Royal Australian Navy (RAN) expressly for the purpose. Now painted haze gray with a hull number, HMAS *Jervis Bay* (named for a former training ship of the RAN) swung its stern ramp over to the pier and quickly began disgorging the troops and vehicles of 3rd Battalion, Royal Australian Regiment. At the bare port facilities, *Jervis Bay* unloaded her vehicles and cargo at a pace that no conventional military transport could match.[3] At her decommissioning from military service in May 2001 (less than two years later), Rear Admiral Geoff Smith, Australia's maritime force commander, lauded the *Jervis Bay*, saying that she "was precisely the vessel we wanted in the East Timor crisis—we needed to transport personnel quickly, reliably and in large numbers."[4] *Jervis Bay* impressed not only Australians, though, because East Timor was not a purely Australian affair. As *Jervis Bay* began her round trips from Darwin to Dili, C-130 Hercules transports from the United States, the United Kingdom, Australia, and New Zealand were bringing in infantry and supplies as well. The catamaran caught the Americans' eye, largely because her cruising speed on the round trips was forty-two knots, faster than that of any ship in U.S. service. On hauls of up to one thousand miles, ships moving at that speed could actually move troops and equipment faster than comparably priced aircraft (such as the C-130), because the aircraft would need to make many round trips to match the line-haul capacity of the ship.

Big Ships, Big Budget Shortfalls

The thought of a small, fast, adaptable, and inexpensive ship was particularly alluring to the U.S. Navy, with so many stations to keep so far from home. Indeed, few other naval forces figure their force structure so strongly in terms of deployability as the U.S. Navy.[5] This does not mean that the Navy had done much hard thinking before 1999 about how to manage that problem. As Ronald O'Rourke of the Congressional Research Service recently noted "the Navy does not show much evidence, at least to outside observers, of having done very much work for years in the area of alternative force architectures."[6] Instead, the Navy rather stumbled in the 1990s trying to determine what it wanted in a variety of ship types and within unforgiving budget constraints.

SUBMARINES

Trying to keep fifty nuclear-powered attack submarines in the fleet was consuming half the Navy's shipbuilding funds. Each new *Virginia*-class submarine, even after negotiated volume discounts, was forecast to cost $2.2 billion—the reactors alone were costing well over $500 million each.[7] Submarines were sexy, and proper navies around the world felt compelled to own them. The Canadian Navy had justified its acquisition of four *Upholder*-class diesel submarines from the United Kingdom in the early 1990s in part on the need to train the Canadian surface fleet in how to hunt submarines. In 2003, the Portuguese Navy claimed that the two Type 209 submarines that it planned to purchase from the German shipbuilder HDW—at a cost of 846 million— would be used to patrol the Portuguese coast. What they would be searching for was not entirely clear.[8] More sensible, perhaps, was the Swedish move to modify its five submarines—HMS *Södermanland*, HMS *Östergötland*, HMS *Götland*, HMS *Uppland*, and HMS *Halland*—with air-independent Sterling engines,[9] divers' locks, and warm water life support systems for intelligence collection and commando operations.[10] The U.S. Navy was attempting to justify the size and cost of its attack submarine fleet as a means of intelligence collection, but collecting signals with a single mast protruding from the water seemed a strange way to spend a few billion dollars.

ARSENAL SHIPS

The Navy never quite emotionally got over the retirement of its large cruisers— ships like the *Long Beach*, *South Carolina*, and *Texas*. In the 1980s, Admiral Joe Metcalfe's extolled the concept of a 20,000-ton Strike Cruiser; in the 1990s, this became the Arsenal Ship, a floating battery of perhaps 500 cruise missiles for over-the-shore bombardment. At one point, the Navy intended to procure six of these.[11] The program was cancelled in 1998 after the contractors actually encouraged the Joint DARPA-Navy program office to terminate it in favor of proceeding with the Surface Combatant (SC)-21 project. The mission—serving as a floating battery of long-range cruise missiles—was eventually taken by four nuclear submarines whose Trident ballistic missiles were replaced on a one-for-seven basis with Tomahawk cruise missiles. This provided fewer missiles, but after the Cruise Missile Diplomacy era of the 1990s faded, fewer were needed. The converted submarines *Ohio*, *Michigan*, *Florida*, and *Georgia* also possessed far greater stealth.[12] The ships were converted with about twenty years of service left, and with dual crews, would spend about fourteen of those years in missile firing positions.[13] Conversion, costing about $400 million per ship, was complete in fiscal year 2006. Each now carries 154 Tomahawk missiles and 66 SEAL commandos with their own mini-submarine.[14]

MULTIPURPOSE MISSILE SHIPS

The aforementioned SC-21 project eventually split into the CVNX project— the design of a new supercarrier to follow the very successful but not-very-updated *Nimitz* class—and the DD-21 project—the design of a new guided missile destroyer. The DD-21 Land Attack Destroyer was cancelled in 2002 by then Defense Secretary Donald Rumsfeld because the design was considered insufficiently advanced for the plans he was laying for the U.S. armed forces. To some extent, however, the program rather morphed into that of the DDX, or experimental destroyer. The DDX, later named the *Zumwalt* class, involved some distinct departures from Navy practice: the Naval Surface Warfare Center in Carderock, Maryland, had been experimenting for thirteen years on tumblehome hull designs for radar avoidance, and this was incorporated into the ship's plans.[15] The ship would also feature new 155-mm guns firing rocket-assisted GPS/INS-guided shells to ranges exceeding 100 kilometers. Modular systems redundancy and automated controls would provide superior survivability under fire.[16] An S-band radar was chosen for better counter-battery fire and medium-range missile defense. The helicopter deck was expected, but the 'in-stride' mine avoidance sonar would be a relatively new feature.[17]

The ship would also famously feature a turbine-electric propulsion system. One of the many benefits of electric drive was a reduction in the required crew, which had become an increasingly important theme in the management of the Navy.[18] Indeed, the *Zumwalt*s would, in the Navy's plan, have crews of only one hundred—and every sailor would have his own stateroom.

As the program evolved, however, the future of the *Zumwalt*s grew unclear. The original plan for a production run of thirty-two ships was cut to just eight in a Pentagon budget decision in 2005. Congressional action in 2006 recommended that just two be built as technology demonstrators. The reasons were clear: the ships were very expensive—over three billions dollars each for the first few—and the need for them was in some dispute. While their long-range guns and cruise missiles could provide a great deal of firepower, firepower was not lacking within the U.S. armed forces. Missile defense might be their true forté, but the need for more than a few missile defense ships—keeping station near Iran and North Korea—was not clear.

VERY EXPENSIVE CARRIERS

The centerpiece of the U.S. Navy, like no other navy, was its nuclear-powered supercarriers.[19] Building one was an enormously complicated undertaking, consuming several billion dollars and roughly seven years. At that low rate of construction, only one yard could be supported: Northrop Grumman Newport News (NGNN), a century-old yard originally known as Newport News Shipbuilding and Drydock Company. As a monopolist with a particularly ugly and unrepeatable integration problem, NGNN was not beloved by the Navy for any measure of efficiency. In the late 1990s, its overhaul of the *Nimitz* had cost $1.5 billion, roughly $250 million more than had been budgeted.[20] In 2002 and 2003, the yard had made such a hash of the overhaul and refueling of the *Eisenhower* that the Navy was finally prompted to stop its customary practice of awarding bonuses for work that was behind schedule and over budget.[21] The Navy claimed to have been promised that Newport News would be able to cut costs further, but its work on the carrier *Ronald Reagan* was also later badly over budget.[22] Worse, its internal communications and advanced networks (ICAN) system aboard the latter ship was unavailable about 25 percent of the time and was sometimes subject to "catastrophic and very disconcerting" failures.[23]

STREETFIGHTERS IN THE LITTORALS

The Navy's commitment to all these large and expensive ships was sharply curtailing its ability to cover a wide selection of points around the globe. The search for solutions was on even before the Navy's commitments increased as the war against political Islamism started in 2001. As U.S. Coast Guard Commandant Admiral James Loy put it in 1997, "with a 600-ship Navy, forty or so Coast Guard cutters were virtually an after-thought. With a Navy of 116 or fewer surface combatants, and in a world plagued with regional instability and strife, however, our forty major cutters, along with several hundred coastal patrol boats, take on new significance."[24]

To fill this gap, the Streetfighter concept was proposed by retired Captain Wayne Hughes of the Naval Postgraduate School in Monterey, California, and retired Vice Admiral Arthur Cebrowski, later director of the Office of Force Transformation at the Pentagon. As Cebrowski put it, the Navy needed a ship that would be well suited to "babysit the petri dish of festering problems we have around the world."[25] The Streetfighter eventually gained programmatic fruition as the Littoral Combat Ship (LCS): a fast, maneuverable, shallow draft vessel of about 2,200 tons (effectively a small frigate) intended to destroy small craft, diesel submarines, and mines.[26] The ship would embark several helicopters and small boats, both manned and unmanned, and would be designed expressly to deliver commando forces ashore.[27] Networked with the forces both further offshore and onshore, the ships would have a cost target of $220 million—one-fifth that of an *Arleigh Burke*–class guided-missile destroyer.

The point of all this, according to the Navy's requirements document, would fundamentally be to guarantee access to and through coastal areas around the world.[28] Design contracts were awarded in July 2003 to General Dynamics' Bath Iron Works, Lockheed Martin's naval systems unit, and Raytheon

Integrated Defense Systems.[29] In late May 2004, the Navy asked GD and Lock-heed Martin to each build two ships as demonstrators. The decision to proceed in parallel with competing designs had not been undertaken since the 1940s.[30] Raytheon's team was praised for its innovative concept, but the Navy was concerned about the management and design skills of the team. An innovative Australian shipbuilder, Austal (see below), actually designed GD's radical welded aluminum trimaran hull, and would assist GD with the construction.[31]

The LCS was a departure for the Navy: its last major experience of littoral warfare had been during the Vietnam War in the 1970s.[32] Indeed, as successful as it had been in dominating the oceans, one could argue that the U.S. Navy had had scant recent experience in recent decades in being shot at. In October 2000, when the destroyer *Cole* was holed and nearly sunk by a suicide boat attack, and seventeen of its complement were killed and forty-two were injured, the crew had to swing fire axes to open the first-aid boxes, as they had been locked shut for better security against pilferage. Such were the priorities of the sea-going Navy.[33] Before the 1940s, though, the Navy had largely been a small-war force with a widely deployed force of frigates (and later cruisers) built with at least some view toward littoral combat.[34]

The problem was that guided-missile destroyers were simply not a cost-effective way of fighting every skirmish in coastal waters. Tracking and boarding merchant and fishing vessels that may be smuggling weapons called for a less costly ship.[35] The size of the problem was huge: 90 percent of maritime cargo moves in containers, but less than 2 percent of all shipping containers are ever inspected before they reach their destinations.[36] An additional problem is that larger warships are not suited by their size to coastal combat. As Captain Mark Busby, commodore of the U.S. Navy's Destroyer Squadron 31 and leader of the effort to interdict Iraqi oil commerce in the Persian Gulf in 2002 put it, "we push as far north as the draft of our ships allows us to go. That's where all the action is."[37] Some small combatants, such as corvettes or frigates, would thus be essential to policing these waters.[38] Besides, the Navy could expect few air defense problems for the foreseeable future, and increasing the size of the fleet given shipbuilding budget constraints (and a shipbuilding budget largely given over to submarines) would simply require more, smaller ships with multiple crews.[39]

Military Experience with Aluminum Catamarans

It was in this context that U.S. officers had begun to look enviously upon the Australians' sleek and exotic catamaran. Pressing requirements led to prompt action. As noted in figure 5.4, Australian and U.S. military experience with catamarans has since involved five ships: HMAS *Jervis Bay*, MV *Westpac Express*, USAV *Spearhead*, USAV *Joint Venture*, and HSV *Swift*.

CHARACTERISTICS OF NOTABLE MILITARY CATAMARANS

Ship	Primary Military User	Length (feet)	Beam (feet)	Draft (feet)	Displacement (tons)	Speed (knots)	Builder
Westpac Express	U.S. Marine Corps	331	87	14	750	35	Austal
Jervis Bay	Royal Australian Navy	284	85	12	1250	45	Incat
Spearhead	U.S. Army	370	100	12	1102	42	Incat
Joint Venture	U.S. Army	370	100	12	1102	42	Incat
Swift	U.S. Navy	316	87	11	940	42	Incat

MV *Westpac Express* was initially used for a year-long trial by the USMC's Third Expeditionary Force to transport entire battalions of troops to and from its main base on Okinawa. The effort was so successful that the Marines kept the ship on contract for years afterwards. The main experience, however, came in two serious military campaigns: the East Timor peacekeeping effort in 1999 and the invasion of Iraq in 2003.

EAST TIMOR, 1999

HMAS *Jervis Bay* was so frequently seen transporting troops from Darwin (in Australia's Northern Territories) to East Timor that she became known as the "Dili Express." The regularity of the trips led to the perception of the ship as a symbol of Australian assistance to the newly independent country.[40] In two years, the *Jervis Bay* completed 107 round trips delivering 20,000 passengers, 430 vehicles, and 5,600 tons of stores to the United Nations Transitional Administration on East Timor (UNTAET). Under Commanders Leslie Vaughn Rixon and Jon Dudley of the RAN, *Jervis Bay* typically completed the 430-nautical mile one-way trip in less than eleven hours. The ship at various times motored with dual and single crews as the Australians investigated different operating concepts. While an excellent experiment, *Jervis Bay* was absolutely needed operationally. The vessel had been chartered because the RAN's older transports had been retired in decrepit condition, and the new vessels, the

Manoora and *Kanimbla,* were still undergoing refitting, as they had recently been acquired secondhand from the U.S. Navy. *Jervis Bay* was also economical: the total lease cost to the Crown was just A$16 million. This was rather favorable compared to the A$60 million that the RAN paid in 1994 for the two U.S. ships, particularly including the A$400 million in repairs and modifications that they subsequently required.[41] When the ships were ready, the *Jervis Bay*, it was thought, could be returned to Incat.[42]

IRAQ, 2003

Three Australian-built catamarans served in the Iraq campaign under U.S. colors. HSV *Joint Venture* was taken into U.S. military service in October 2001, after installation of a helicopter deck, stern-quarter ramp, rigid-hull inflatable boat deployment gantry, troop facilities and additional crew accommodations at the Incat shipyard in Hobart.[43] During the Iraq campaign, she served as a floating base for the SEALs who seized Iraqi oilrigs in the northern Persian Gulf at the start of the war. USAV *Spearhead*, used to ferry soldiers and equipment around the Middle East, impressed the Army with its cruising speed of forty-nine knots.[44] HSV *Swift* joined the U.S. Fifth Fleet in August 2003, primarily to serve as a high-speed logistics vessel. Unlike the other catamarans, *Swift* was a military project from the beginning, so the interior spaces were built out in a much more militarily suitable way than would be expected from a converted car ferry.[45] On the other hand, the Navy did some strange things initially, such as insisting on nonskid paint on the flight deck instead of Incat's far more durable (and equally effective) roughed aluminum surface.[46] The Army was less attached to military convention: leaving the rust-proof aluminum unpainted on its vessels saves nine tons of weight, which can be used to haul cargo or save fuel.[47]

Requirements and Developmental Antecedents

The Army's motivations for using the catamarans became apparent in the Persian Gulf. The Army operates an efficient and significant fleet of transport ships primarily for ferrying troops between ports in combat theaters. The lightweight, twin-hulled Australian ships could carry more, faster, further, through water with more than twenty-five-foot waves and to far more ports than the vessels on which the Army had expected to rely for that purpose.[48] The Army quickly established a program to acquire fourteen of the 1,100-ton catamarans as Theater Support Vessels (TSVs) by 2012, and awarded the initial contract to a team composed of Incat and Bollinger, a small shipbuilder in Louisiana and Texas noted for its construction of patrol boats and small cutters for the U.S. Coast Guard. Assuming that two of the TSVs would be undergoing maintenance or refitting at any given time, the other twelve could move an entire Stryker brigade four hundred miles in a single squadron sortie. As noted above, at this distance sealift was more effective than airlift: the same movement by aircraft would require 254 sorties by C-17s.[49]

As the Navy expected its catamarans to undertake a wide range of naval missions, such as landing commandos, hunting mines, and delivering boarding parties, its requirements for a High Speed Vessel (HSV) were somewhat more expansive, though still straightforward:[50]

- optimal manning with a permanent crew of twenty, augmented with staff as necessary for tailored specific mission profiles, and berthing for a crew of fifty-one;
- galley for permanent crew messing, installed seating for 150, capacity for as many as 300 passengers, a crew lounge, multipurpose spaces for administration, planning, and berthing, and a surgery;
- A flight deck permitting simultaneous launch and recovery of two SH/MH/CH-60S SeaHawk helicopters, plus helicopter equipment storage lockers, a multifunctional elevator, capacity for movement of helicopters from flight deck to and from the vehicle deck, and movement of vertically replenished cargo from the flight deck to and from the vehicle deck;
- 28.740 square feet of cargo space on the vehicle deck;
- launch and recovery of small boats and amphibious vehicles through a "moon pool" in the vehicle deck;
- self-defense capability with Raytheon's proven Rolling Airframe Missile and netted to the fleet-wide Cooperative Engagement Capability;
- 1,100 ton hull driven by four diesel engines and four waterjets.

—all the expected communications equipment of any U.S. Navy combatant.

Despite the success of these vessels, it must be admitted that the Australian catamarans were not the only source of ideas. As the Army and the Navy began to think increasingly about small ships—whether TSVs, HSVs, or LCSs—the services drew upon the knowledge of a variety of smaller navies with modern and innovative ship designs:[51]

KNM SKJOLD

In October 2001, the Royal Norwegian Navy sent one of the most innovative naval vessels in the world to the United States to assist with the suddenly expanded requirement for port security in North America. The vessel and her crew remained until the following September and were the subject of intense interest (and testing) by the U.S. Navy and Coast Guard. KNM *Skjold* is a 260-ton surface effect ship built of glass- and fiberglass-reinforced plastic and a variety of other composite materials. Despite her small size she carries a 76-mm Oto Melara gun and eight anti-ship missiles with a range of 100 kilometers. What is more impressive is her speed and economy: with a crew of only fifteen, she makes fifty-five knots in calm seas (with admittedly poor endurance), and at least fifteen knots in sea state 6. The ship's technology was the basis for Raytheon's bid in the LCS program. The U.S. Navy, however, was quite concerned about Raytheon's ability to increase the scale of a surface effect ship by a factor of roughly ten, especially as Raytheon is not actually a shipbuilder.

HMH *THETIS*

The Royal Danish Navy's *Thetis*-class Stanflex 300 (standardized, flexible) corvettes illustrate how far a ship designer can take modularity. The 320-ton, 54-meter ships each can mount up to four modules with a variety of payloads: surface-to-surface missiles, surface-to-air missiles, guns, anti-submarine warfare equipment, variable depth sonars, mine counter-measures equipment, cranes, oceanographic and hydrographic research equipment, pollution fighting gear, and signals intelligence gear.[52] At its mightiest, the Stanflex 300 displaces nearly 500 tons and carries eight Harpoon surface-to-surface missiles, six Sea Sparrow anti-aircraft missiles, two wire guided torpedoes, a 76-mm gun, and a soft-kill weapon system for protection against incoming missiles. The larger Stanflex 3000 ships are the same size as a US Coast Guard High Endurance Cutter and perform the same mission, but rather than a crew of 175, they carry one of only sixty.[53]

HMS *VISBY*

Another composite-hulled design of note was the Royal Swedish Navy's *Visby* class of stealth corvettes. Primarily designed for hunting mines and submarines, the *Visby* was also built with a strong capacity for destroying surface ships. The design is thus quite modular: each ship can take a helicopter or, in lieu of the hangar and elevator, a vertical launcher with eight medium-range surface-to-air missiles. The vessels carry (according to the Swedish pattern of manning) a complement of twenty-one officers and twenty conscripts. What is most notable about the vessels is the exquisite attention paid to reducing their signatures in all aspects. Combined diesel or gas turbine (CODOG) engines drive waterjets, which were chosen primarily for quieting. Carbon fiber over foam construction improves emissions control by inherently interfering with outgoing radio signals. The entire hull and superstructure of the ship progressively tumbles home away from the gunwales to scatter incoming radar energy.

The research program for these stealth corvettes was started in 1988 after Soviet submarines had been found routinely violating Swedish territorial waters. Extensive simulation and wargaming had indicated that stealth would be a more cost-effective defense than a heavy suite of defense weaponry and countermeasures. As Kockums puts it in its promotional literature, "invisibility is more likely than invincibility."[54] In selecting technology for the vessels, the Swedes searched far and wide. The diesels were procured from MTU, and the gas turbines from Vericor, its U.S. subsidiary. Rolls Royce supplied its Kamewa waterjets. The sonar suite, including the towed-array (an unusual feature in a 600-ton vessel) was built by Computing Devices Canada.[55]

Kockums, which first launched a ship for the Swedish Crown in 1682, employs 1,200 people at facilities in Malmö (submarines), Karlskrona (surface vessels), and Muskö (fleet maintenance); and is itself today a division of the German industrial conglomerate ThyssenKrupp. HMS *Visby* entered service in 2005; the next four vessels following at six-month intervals, with the last, HMS *Uddevalla*, coming operational in 2007. Kockums also extensively upgraded the corvettes *Malmö* and *Stockholm* with new superstructures featuring lower radar cross sections, new engines, and new combat systems. The company has also developed a range of "Visby Plus" designs, up to 1,500 tons, for foreign customers who would like a larger stealth corvette.[56] Indeed, Lockheed Martin chose a semi-planing monohull similar to Kockums' approach as the basis for its LCS design because it was much more fuel-efficient at low speeds than a surface-effect hull.[57]

Problems with Aluminum Catamarans in Military Service

Despite these innovations, naval conservatism is almost tautological. There are, after all, very good reasons for hesitancy in adopting new concepts for traveling the seas. In the case of the catamarans, the first point of resistance was the aluminum hulls. Notably, aluminum was never considered for the design of the Norwegian and Swedish missile ships, for three reasons. First, aluminum is thought not to hold up well structurally in intense fires.[58] For this, it had garnered a bad reputation: a spectacular fire had destroyed the aluminum superstructure of the frigate HMS *Sheffield* in the Falklands War, resulting in the total loss of the ship. Second, it is difficult to weld, though the two primary Australian builders of aluminum catamarans—Austal and Incat (see below)—had invested a great deal of time and money in developing better and lower-cost techniques for accomplishing this. Third, even if the welds can be well done, they do not hold up well under sustained fatigue.[59] For a car ferry, which should be retired from service after twenty years, this is not such a problem. For a military vessel that might be expected to serve for thirty of more, this could be problematic.

The second issue concerned their ride: seasickness aboard catamarans can be a problem in high seas. Although the ships are very comfortable in reasonably calm seas, they are notably uncomfortable in rough ones. The twin hulls compensate for the shallow draft by making the ship stable, but the stability brings a very stiff ride, with considerable slamming against the mid-hull when pitching.[60]

Finally, in the U.S. Navy's case, one of the primary attractions of the ships—high-speed mine warfare—was finding other outlets. Mine warfare is indeed important—since the end of WW2, fourteen of the seventeen U.S. Navy ships damaged by enemy action have been hurt by mines. It is also an area that the U.S. Navy has chronically underfunded. It was thus unsurprising that HSV *Swift* was used in part in the Persian Gulf as a minehunting command vessel, controlling robotic submarines and dive teams looking for sunken and moored explosives. The eventual problem was that the HSV *Swift* garnered less interest in the U.S. Navy as a minehunter, because the Navy's new plan for mine hunting focuses on delivering in-stride (organic) network-enabled mine avoidance systems to all combatants. As noted above, individual frigates and cruisers will increasingly carry their own mine-hunting remote-controlled submarines for this purpose.[61]

Incat

The firm responsible for four of these five vessels was Incat—International Catamarans of Hobart, Tasmania. Incat was a small firm by shipbuilding standards: at its peak in 1996, the company had just 1,000 employees and 300 subcontractors, but at that time was the state's largest private employer. The company was founded in 1975 when the Tasman Bridge over the Derwent River in Hobart collapsed after a ship rammed it. Local boat builder Robert Clifford immediately established a ferry business that eventually hauled nine million passengers across the river over the next two years.[62]

Clifford's designs were innovative, and his ships won attention for their capabilities. In 1990, Incat's first fast ship, the *Hoverspeed Great Britain*, won the famous Hales Trophy for breaking the transatlantic passenger service speed record. In 1998, the *Catalonia* and the *Cat-Link V* set new records in

June and July, respectively.[63] The firm's capabilities grew over time within the market segment that he had selected. While the largest aluminum catamaran ever constructed in the United States was a 50-meter vessel, Incat had facilities and plans for building multihulls up to 120 meters long. Aluminum was an excellent choice for ferries, as its light weight translated to greater lifting capacity and higher speeds (or better fuel economy at the same speed). Incat catamarans were usually equipped with diesels for economy, but with gas turbines they could exceed fifty knots. Selling ferries at A$60–100 million each; and competing against yards in Scandinavia, Italy, and Spain, Incat and its competitor Austal amassed a 40 percent share of the market for high-speed ferries worldwide.[64]

This does not mean that the company was universally successful at all times. Shipbuilding can be a cyclical market, and the general worldwide economic slowdown in the late 1990s severely dented demand for new car ferries. The company's practice of speculatively building ships, chartering them to operators, and retaining ownership led to a rather heavy balance sheet. In April 2001, labor demands for higher pay and a 38-hour workweek led to a partial twenty-four hour strike. Oddly, this was while the company was losing large sums with several completed ships tied up on the Derwent River awaiting buyers or charters.[65] In July 2001, fortunes began to turn when the Pentagon leased a 96-meter catamaran—the *Joint Venture*—for two years for the impressive fee of A$50 million. Up to this point, the company had sacked 100 employees, and competitor Austal (in Western Australia) had sacked 200.[66]

Some rough patches remained ahead. From March 2002 to February 2003, three of the six companies in the Incat group were in receivership—Clifford's firms at that point owed A$70 million to the National Australia Bank and A$30 million to the Tasmanian state government, and held financial interests in fourteen of the sixty Incat-built ships afloat. After failing to sell a single ship for over a year, the company sacked most of the remaining staff. All the same, the managing director, Clifford's son Craig, stayed on throughout the receivership since his bankers had fundamental confidence in the management and the long-term forecast that the market would turn.[67] There were reasons for this conviction. Clifford told an interviewer during the invasion of Iraq that the U.S. military alone could use as many as one hundred high-speed catamarans, which would lead to revenues to Incat of roughly A$1 billion and a work force of at least 5,000 in Australia and the United States.[68] While a rather round figure, one could imagine such a fleet—the *Oliver Hazard Perry*-class frigates displaced almost four times as much, but some seventy-one have been built for the U.S., Royal Australian, Spanish, and Taiwanese navies over three decades.

Austal

One of the five catamarans in question–MV *Westpac Express*—was built by Austal (Australian Aluminum Shipbuilding), Incat's rival on the other side of Australia. The shipyard is situated on Jervoise Bay in Henderson, Western Australia, about eight kilometers south of Freemantle. Its production facilities have some 20,000 square meters of covered production space, which allows the company to build two 100-meter vessels and four 40-meter vessels simultaneously. Austal's vessels are comparably outfitted for small crews and slight administration, with automated machinery monitoring and computerized equipment manuals and electrical drawings.

Like Incat, Austal was struggling financially just as its products were gaining acclaim. In the fiscal year through the end of June 2003, Austal lost A$18.7 million on revenue of A$307.8 million. The revenue figure was down 10.4% from the preceding year. Problems included higher than expected labor costs due to new and inexperienced staff, high marketing costs in tackling the military market, and delays and rework due to inexperience with classification requirements in the United States. The firm had previous experience building a class of eight monohull welded aluminum patrol boats for the Australian Customs Service (seven, interestingly, are named after bays on the Australian coasts). This experience earned Austal the role of constructing the first two General Dynamics LCSs at the Australian firm's new U.S. yard in Mobile, Alabama. After all, Austral was already building a 127-meter fast ferry with the same trimaran hull form at its Australian facilities.

Small Warships from Small Shipyards: The Catamarans as a Case Study

Reviewing the criteria laid out in chapter 3, Incat and Austal's success in adapting its car ferries to military purposes is very understandable. The ships are excellent examples of the type of system in which small arms contractors can be quite successful vis-à-vis their larger competitors.

INNOVATION WITHOUT R&D INTENSITY

Thousand-ton catamaran transports were a serious innovation for most navies that adopted them. The end of the Cold War erased the Soviet Red Banner Northern Fleet as the primary threat to maritime security, but this was replaced by a wide variety of low-level threats. For some fleets in Europe— the Danes, the Belgians, and the Norwegians, for example—this suggested that larger vessels were needed for ocean-going activities far from home on stability and support operations, if not force protection. For most other fleets, this called into question the need that all new frigates and destroyers displace 6,000 tons or more and carry high-altitude air defense systems. Despite the economies of large ships, more ships can cover more water. Further, bridge and engine room automation (if belatedly adopted by the U.S. Navy) had by the 1990s reduced crew sizes so that larger numbers of smaller vessels were relatively affordable.

At the same time, these ships were not science projects—the fundamental concept of the aluminum catamaran ferry was thirty years old by the time the RAN put the first military version on the Dili run. This meant that firms like Austal and Incat could forge ahead in the industry without the huge financial risks of vast R&D outlays. At Incat, the R&D department was roughly twelve people working on improved fabrication techniques, weight reduction, improved propulsion systems, better power management, and improved lift from the underwater wings.[69]

The yards also had ample room for additional innovation. The broad beam, high speed, and high stability of catamarans had led to suggestions for a wide range of military concepts. Since the twin hulls preserve a high degree of hydrodynamic efficiency in an otherwise very beamy ship, the extreme width can be used as a stable platform for a variety of uses. In 2003, Incat put forth plans for a "Sea Frame" catamaran that could be used for a wide range of military purposes. The 113-meter ship would displace 1,500 tons empty, and require a complement of just thirty. The vessel could handle troop transport (carrying 730 troops on short hauls), amphibious assault (with a helicopter pad for a CH-47, a large hangar for vehicles and helicopters, and 350 troops), or air support (carrying a few F-35B Joint Strike Fighters, several helicopters, rigid inflatable boats, a few armored cars, and 350 troops).[70] The ample space aboard the catamarans would mean that installation of future unexpected mission equipment could be relatively easily accomplished.

Thus, the value of the ships would probably hold over time, most likely to navies, and certainly to the contractors who would continue to produce and upgrade them. As Vice Admiral John Nathman, the U.S. Navy's Deputy Chief of Naval Operations for Warfare Requirements and Programs, admitted in April 2003 to the House Armed Services Committee, the Navy had chosen a small ship with a wide beam as its option for coastal warfare intuitively. The service had not bothered to consider, say, helicopters on larger ships further from the coast, though he did promise that the serious analysis would come later.[71] The analyses were foregone largely so that the Navy could just do something, and experiment its way through.[72] Navies the world around expected an enduring need for ships, and would buy more reflexively. Hence, in the LCS program, the Navy decided in June 2004 to buy two entirely different small frigate designs, with two examples each from Lockheed Martin and General Dynamics.

SKILL-INTENSIVE PRODUCTION

As Lance Balcombe, Incat's chief financial officer put it,

> [Our] infrastructure requirements are quite simple. All we require is a large shed, welding equipment and hand tools. We have a production technique where we roll the ships down a line that has three positions. Allied to this we have a large prefabrication area where we plug larger components into the vessels on the production line. The third position on the line has a dry dock so we can float the ship inside the shed and do all our engine and machinery alignments in there.[73]

In short, catamarans are economical ships, but their production is not akin to that of economy cars. Shipbuilding is one of the most skilled of the heavy industrial trades—even more so than automotive assembly, but its low production rates limit its capital intensity. Based in Hobart, Tasmania, Incat has some particular advantages in a low-cost labor force and an excellent climate (temperate with low humidity) for shipbuilding. As a small and skilled firm, Incat is fortunately in a rather heterogeneous industry—one in which smaller firms can burrow themselves into viable niches. Ship repair is a relatively fragmented industry worldwide, and may remain so. In the United States, for example, only three factories in 2004 were producing fighter aircraft for the USAF, U.S. Navy, and USMC. The Navy, however was buying ships from six different major shipyards. In European NATO, about twenty shipyards were operating, even if that number might have been twice as large as needed.[74]

MEDIUM-SPEED LEARNING

In its first thirty years, Incat built sixty ships. That the company describes each one was an improvement on the previous rather indicates the speed of learning in the industry. Construction rates are low, but plenty of opportunity exists in each project for adjustments to processes and parts of the product. This places shipbuilding far from the dynamics of the armored combat vehicle business, in which production is sometimes closer to two units per day than two units per year.

At the same time, there was room to slip into the military side of the market. In the United States, for example, production runs were forecast to shorten considerably from the experiences of the *Oliver Hazard Perry* and *Arleigh Burke* classes. The *San Antonio* class of assault ships was planned for just eight or nine ships, and the *Zumwalt* class of destroyers for as few as two. This threw into serious doubt the economics of the larger U.S. yards that were not, in any case, competitive to begin with. While two-thirds of the $11 billion spent on shipbuilding and repair in the U.S. comes from military contracts, U.S. yards have a global market share of less than 1 percent in the construction of vessels of deadweight exceeding 1,000 tons.[75] In looking to smaller yards, as Al Bernard, director of government relations for Marinette Marine put it, "the Navy is going to find an industrial base that really hasn't been tapped into for quite a while, and see that competitive shipbuilding does exist in this country."[76] Marinette, based on Lake Superior in northwestern Wisconsin (and far from any substantive naval base), had built quite a business building small cutters for the Coast Guard, and was seeking work in the LCS program. That is, small shipyards with strongly differentiated products have in many cases thrived even as larger firms struggled in the post–Cold War naval construction downturn.

Big Ships or Small, Big Systems—The Issue of Systems Integration

The ability of a small shipyard to make a difference in naval construction is at once simplified and complicated by the question of systems integration aboard naval vessels. To some extent, Incat and Austal's catamarans were mere platforms onto which payloads could be installed, perhaps in a modular way. As the U.S. Navy convinced itself in the Arsenal Ship and LCS programs, if the installation requires only a relatively loose integration with the rest of the vessel, then the prime contractor for the mission system need not be the prime contractor for the ship. At that point, the program manager can structure the contract in a variety of ways, but certainly need not expect a 300-person shipyard to manage the acquisition of the fire control system.

On the other hand, the complexity of the work limits the applicability of this model. Ships are inherently complex products: in the construction of even a relatively simple vessel, the shipbuilder must control the scheduling of work and materials moving within the physical constraints of tight spaces. The model has also not had a uniformly happy history. At present, aircraft carrier combat systems in the U.S. Navy (whether for the *Nimitz*, *Tarawa*, or *Wasp* classes) are amalgamations (if relatively effective ones) of different subsystems installed without a governing vision a priori. Interoperability is a huge issue, and it is complicated by constant software upgrades to meet every new surmised threat. In February 1998, two Aegis missile ships in San Diego failed to get underway with their battle group due to interoperability failures among their installed combat systems.[77]

This concern over the ability of contractors to manage complex intraplatform and interplatform integration problems has already had an impact in contract awards. Northrop Grumman beat General Dynamics to the DDX contract largely because the Navy was impressed by Northrop's ability to integrate a wide variety of subsystems without overreliance on suppliers. As noted in chapter 4, this had plagued Northrop Grumman's space satellite business before the acquisition of TRW, and Litton's shipbuilding divisions were not known for electronic systems knowledge before they were integrated into Northrop.[78] In the short run, it could be seen as a relative vindication of Northrop's strategy of maintaining a balance between vertical integration and specialized subcontracting.

At the same time, the Navy rather rushed into the LCS program, which could be thought to have exacerbated the technical risk of the program with respect to just those integration problems. There were, on the other hand, reasons to proceed at full speed. As Admiral Cebrowski put it, the Pentagon frequently slows down programs "to mitigate technical risk. When we do that, we aggravate every other kind of risk that we have. We aggravate financial risk, customer risk, requirements risk . . . it's not what you want to do."[79] The LCS program, truth be told, basically was to produce a fleet of small frigates with corvette-sized crews, novel hull forms, and modular weapons packages. As noted above, ships with these features were already common in Europe, so the program presented less technical risk than might be supposed.[80]

Big Yards, Small Yards—The Question of Alliances in Shipbuilding

In preparing for the LCS competition, General Dynamics (GD) adhered to its previous strategy but took a more radical turn with it. In bidding on the contract, Lockheed selected primarily specialty shipbuilders as its teammates—this is unsurprising for a firm with such a well-respected naval electronics business. GD, on the other hand, invited two large rivals onto its team: Northrop Grumman and BAE Systems. More dramatically, it asked Austal, a rather small shipbuilder, and one of the world's specialists in aluminum multihull construction, to build the first two units of its offering. This turn of strategy is even more interesting in that GD is a shipbuilder (as the owner of the Bath Iron Works in Maine and the Electric Boat Company in Connecticut and Rhode Island), while Lockheed Martin is not. As one financial analyst put it, "General Dynamics has no qualms about cooperating with the competition if that is what needs to be done to win contracts."[81]

This dynamic of alternative cooperation and competition—prevalent, frankly, in the arms industry for some time—was further shown in the U.S. Coast Guard's (USCG) Deepwater cutter competition. Litton Industries' Avondale yard proposed its own solution to the USCG's requirement, but Litton's Ingalls yard and two other Litton divisions served as members of Lockheed Martin's winning team. After all, cooperating with the competition (short of colluding, of course) limited the downside risk of losing. Deepwater was forecast to be a $60 billion project, so even with the possibility of a significant budget reduction, there was plenty of money to go around, whoever won.[82]

While alliance building thus makes sense for a large firm like GD, the question must still be asked in the reverse: does the Incat and Austal's approach to marketing their ships overseas through alliances make sense? A small firm can easily find itself on the short end of such a deal. In this case, however, the answer is generally affirmative. Reviewing the criteria laid out in chapter 3, one finds good reasons to proceed:

CONSIDERABLE CHANGE WITH RESPECT TO PROCESSES AND GOALS

Most navies—and the U.S. Navy in particular—have had only recent experience with aluminum multihulled vessels. The LCS and TSV programs were still experimental in 2004. Multihulls of these types were being proposed for all sorts of roles, from mine hunting to shore bombardment to carrying commandos and assault helicopters. As suggested above in the discussion on innovation, some the best uses for military catamarans were yet to be discovered. At the same time, the question of which uses would be in the greatest need was not clear, and probably would remain unclear for some time in the new international security environment of the twenty-first century. As suggested at the beginning of this chapter, many navies—the United States in particular—were having some difficulty in the new environment deciding exactly what they want in warships. The task of developing new ideas—and then newer ideas—for using these novel ship types would be shared between these navies and their shipbuilders. While a larger shipyard might be more capable of driving down costs over the course of a large class of ships, a smaller firm could arguably respond to the emergent needs of its customers faster on each of the ships. Building essentially only one type of ship with many variations, Incat could be expected to defend its franchise most aggressively with alacritous customer service.

MODERATELY LEAKY KNOWLEDGE

The high-speed catamarans of Australia did represent a rather different approach to shipbuilding and one that was not too readily learned. Welding aluminum is quite different from welding steel, and building catamarans is quite different from building monohulls. Incat had successfully licensed its designs and technologies in the past without losing its position as the leading builder of high-speed aluminum catamaran ferries. Still, the work is complicated, so an arms-length licensing involving the sale of blueprints would not quite be enough. Instead, Austal and Incat would need to supply a small but significant number of skilled engineering and design staff over to GD and Bollinger, respectively, to market their ships in large numbers to buyers concerned about domestic content (such as the U.S. Army and Navy).[83] On the other hand, acquiring or merging with Austal or Incat might not be worth the trouble—Australia's geographical remoteness would make cost-reduction through services consolidation problematic. Worse, commercial shipbuilding had never been in GD's set of skills. For its part, Bollinger had primarily built, in addition to patrol boats, small utility craft like tugs, fishing vessels, dredges, and drilling rigs, so the fit with car ferries was not obvious.

MODERATE POTENTIAL FOR A SHAKEDOWN

Here, as before, we consider the ability of the larger contractor to shake down the smaller contractor. If the smaller firm's production assets are too specific to the end product, then vertical integration tends to make more sense than a contracting relationship. If the assets are quite fungible among different markets, then an open purchase of the products from the assembly line tends to make more sense. In the case of shipbuilding, it is true that shipyards build ships, but certain docks, assembly houses, and shipbuilding crews build certain types of ships. Catamaran car ferries, it should be noted, are neither rigid inflatable boasts nor aircraft carriers. The assets required to produce them are not totally adaptable to other markets—note how Incat and Austal struggled in the late 1990s. On the other hand, the shipyards are not one-of-a-kind. While just two navies in the world have ever bought even one nuclear-powered aircraft carrier, plenty could be considered in the market for thousand-ton catamarans; if those markets did not pan out, then the same ships could be modified for commercial service, and back again (see below). Thus, since the production technologies involved were only somewhat expropriable, a long-term alliance was a viable solution for a small firm like an Incat.

Future Possibilities for Aluminum Catamarans

Catamarans today are notable not just by their utility, but by the lack of a consensus around their role. While the RAN leased the *Jervis Bay* for two years, it has no present written requirement for such a vessel today, and is instead fulfilling the troop transport and landing mission with HM Australian Ships *Kanimbla, Manoora*, and *Tobruk*. All the same, Her Majesty's government in Australia lobbied Washington on behalf of General Dynamics' bid in conjunction with Austal for the Littoral Combat Ship. The Australian Ministry of Defence had formed a team to push Australian companies as potential subcontractors in the LCS program, specifically marketing them as offering top-flight technology with much lower labor rates than found in the United States. This does not necessarily indicate that the government in Canberra is not interested in military restructuring, but simply that it has other priorities, or thinks that a lease will work again in the future.[84] Refitted aluminum catamarans (relatively new, and recently in service in British Columbia's George Straits) have also been proposed to satisfy part of the Canadian Navy's requirement for afloat logistics and over-the-shore support for expeditionary units of the Army and Air Force (see the artist's conception below). Even the Chinese Navy was getting into the act, having built a wave-piercing, diesel-powered catamaran in Shanghai in 2004, though probably simply as a technology demonstrator.[85]

In contrast, it was easy to come to a consensus about the role for large missile ships, but the marginal value of larger ones in the fleet was less obvious. The U.S. Navy was making the case for both, though not with even arguments.[86] In 2004, the air defenses of the U.S. fleet were close to impenetrable—in addition to all its fighter squadrons, the Navy had twenty-seven *Ticonderoga*-class and forty-three *Arleigh Burke*-class guided-missile ships armed with the unsurpassed Aegis combat system. This was enough to provide three of four escorts (in addition to submarine-hunting ships of the *Oliver Hazard Perry* classes) for each of the Navy's twelve supercarriers and twelve jump-jet and helicopter carriers. What the Navy lacked were small, expendable vessels to handle the mundane tasks of sea control. Building a 100-meter catamaran would take eight to nine months—far less than the time required to build even a frigate—and this responsiveness would be useful to fleets in future wars. Thus, even as late as April 2004, the Navy was considering pre-empting the LCS competition and directing that the HSVs built on the *Swift* pattern be the solution.[87] The Navy certainly investigated the option: from September 2003 to May 2004, U.S. Navy and Incat mariners and engineers put the *Swift* through a series of voyages and heavy weather trials across the

Pacific, the Indian Ocean, the South Atlantic, the Gulf of Mexico, the North Atlantic, the North Sea, and the Norwegian Sea. On the ship's return to Little Creek, Virginia, the team found minor damage to the superstructure, but none to the hull.[88]

Ultimately, *Jervis Bay* provided an excellent and outgoing example of the adaptability of the catamaran. In May 2004, the ship was put back in service by Speed Ferries on the Dover-Boulogne run across the English Channel. Rechristened *Speed One*, the ship was carrying up to 200 cars and 800 passengers at over 40 knots. Speed Ferries positions itself as an underdog and upstart, lodging complaints with the EU competition authority and the British Office of Fair Trade against the allegedly underhanded tactics of its larger, more established—and in the company's words, "piratical"—competitors.

6

Mountains Miles Apart
PowerScene, the Dayton Peace Talks, and the Demise
of Cambridge Research Associates

The fidelity of this thing is outstanding. It's as good as you would see in real life. We're using it on every mission we can...We come at some targets at 600 to 700 miles an hour. We have seconds to identify the target or we don't drop. PowerScene helps us recognize the target faster and gives the angle we'll be looking from. That can make the difference between dropping and not dropping. The bottom line is, a higher percentage of our bomb runs have been successful.[1]

Background: War and Peace Settlement in Bosnia

The 1991 war against Iraq was the first in which mission planning and rehearsal systems (MP&RS) were intensely used. MP&RS offer two main advantages to air arms: improved bombing accuracy and reduced collateral damage.[2] U.S. forces in 1991 employed two systems: the USAF's Mission Support System II (MSS II), which was used by F-15E Strike Eagle, F-16 Fighting Falcon, and F-117 Nighthawk crews; and the USMC's Map, Operator, and Maintenance Station (MOMS), which was used by AV-8B Harrier pilots.[3] In the campaign, the MP&RS ran on laptop computers and generated plans, routes, and maps for flight crews and commandos. The hardware and software could generally produce images of target areas similar to those that would appear on radar screens in less than a minute. This was not simple programming: the software would need to show the locations of air defenses, define critical sun angles, display clouds and haze realistically, and allow users to select particular vantage points from which better to understand their missions.[4]

To improve on the relatively simple displays of MP&RS that used 1980s computing technology, the U.S. Navy subsequently launched a program to combine digital maps, ortho-rectified imagery, and digital terrain elevation data to produce three-dimensional visualizations of terrain. The contract was awarded to a small information technology company in McLean, Virginia, by the geographically inappropriate name Cambridge Research Associates. Cambridge was, at the time, a roughly thirty-person firm that was completely focused on computer systems work for the Pentagon. The firm was quite effective at securing some of the most enticing "crumbs"—in particular, the most innovative recombinative applications of off-the-shelf information technology—in large contracts that were undertaken by much larger firms.[5] In the case of PowerScene, Cambridge secured the entire contract for itself.

The results were impressive: PowerScene provided "representative terrain texture, geo-specific terrain texture, synthetic texture, and flat or smooth shaded polygons." Users could "invoke reduced fidelity for portions of the scene that lie outside the region of interest or that lie away from the user's direction of gaze" in order to improve rendering accuracy by the computer, which was dealing with early-1990s processing speeds.[6] Aircrews assigned fixed targets to attack could rehearse their missions by "flying" the routes over a simulation of the terrain in Bosnia, complete with representations of the buildings, roads, bridges, and air defense threat envelopes below. As noted, quick recognition of a target—or a nontarget—could mean the difference between a successful strike and a scrubbed mission, or worse. No similar system was available in the mid-1990s to, for example, Russian pilots bombing rebels in Chechnya, which contributed to the higher rate of unintended casualties in that war.[7]

PowerScene's finest hour, however, was found in the subsequent peace and quasi-partition negotiations. At a meeting in Geneva on 8 September 1995, the various parties to the territorial dispute over Bosnia & Herzegovina agreed in principle to a cease-fire on the basis of a 51 to 49 percent split in territory between a confederation of Bosnians and Herzegovinan Croats, and the Serbian Republic of eastern Bosnia. Proximity peace talks convened at Wright-Patterson Air Force Base in Ohio on 1 November, under the tutelage of Assistant Secretary of State Richard Holbrooke and his military assistant, Lieutenant General Wesley Clark so that the belligerents could work out the details for a permanent territorial and political settlement. Winston Churchill used to brag about having drawn Jordan's borders on a map one afternoon,[8] but the details of the maps at Dayton would be rather complicated. Large parts of Bosnia are both reasonably densely populated and rugged, so defining the interentity border between the Serb Republic and the Bosnian-Croat Federation would be more complicated that drawing a parallel between British Columbia and Washington State. Accordingly, Major General Philip Nuber, of the USAF and the U.S. Defense Mapping Agency, was sent with a team of fifty-five government and contractor staff to support the talks.[9] To support the talks, the team brought a large suite of computing equipment built around Cambridge Research Associates' PowerScene system, including:

- two PowerScene terrain visualization systems, based on PowerScene software and Silicon Graphics workstations, one of which was rigged for large-screen projection,
- three Sun workstations equipped with the ARC/INFO geographic information system,

- four Digital Topographic Support System–Multi-spectral Image Processor workstations equipped with ERDAS Imagine software,
- ten Hewlett-Packard 650C plotters,
- one 3M Remote Replication System (RRS),
- three Canon Bubblejet A-1 copiers, and two more off-base for surge requirements, and
- one large-format Canon Bubblejet 2436 copier.[10]

The team included staff from the DMA's Defense Mapping School for map production and distribution, the U.S. Army's Topographic Engineering Center for engineering and ARC/INFO support, Cambridge Research Associates for supporting PowerScene, 3M for supporting its RRS, and Camber Corporation for ARC/INFO support.[11] Over the next several weeks, the negotiators were up at all hours discussing boundary adjustments. Hubert's mapping team printed more than 30,000 maps for annotation by the negotiators, producing as many as 600 per day.[12] At one point, the team assembled more than one hundred 1:50000-scale maps into a floor-to-ceiling display of the entire country. The most dramatic part of the mapping effort, however, centered on the negotiators' use of PowerScene. PowerScene allowed negotiators to simulate flying through mountain valleys, down roads, and around towns to see how proposed border adjustments would affect the people on whose behalf they purported to negotiate. As Vic Kuchar of the DMA put it, the former Yugoslavs "were totally blown away: they had never seen anything like it."[13] Holbrooke noted that while the room containing the computer equipment was theoretically restricted to those with the appropriate security clearances, it somehow always attracted a stream of visitors who found the fly-throughs more diverting that the bar scene in Dayton.[14]

Every adjustment in PowerScene had to be digitally transferred to a 1:50000-scale DMA map, then digitally rendered as a complex set of polygons so that the DMA team could calculate the shift in areas. Since the very basis of the cease-fire was contingent upon a 51/49 division of territory, this number had to be consistently preserved through all the negotiations. Changes were plotted with a ten-minute delay for rerendering, and with 0.003 percent margin of error in area. At the same time, the team continued to use PowerScene to print (on orders of Holbrooke and Clark) a vast flow of paper maps, in part to maintain the appearance of authenticity and accuracy for the negotiators.[15]

Some tense moments were resolved by reference to PowerScene. When President Slobodan Milošević loudly insisted that the Bosnians' corridor between Sarajevo and Gorazde could not be more than two miles wide, General Clark retorted that the valley through which the road ran was much wider, and that Serbian-held hills on either side of the corridor would leave it too vulnerable to interdiction. When Milošević doubted the geography, Clark led him in a simulated flight down the valley over a round of scotch, noting that "God did not put the mountains two miles apart."[16] Incidentally, widespread use of PowerScene was also useful for convincing the former belligerents that NATO had excellent information about their terrain and facilities, and thus could very quickly plan accurate air strikes again if the discussions fell apart.[17]

The final settlement was signed on 10 October 1995, after only a little more than one month of negotiations. That such a complex division of population, territory, and economic infrastructure could be accomplished in such a short period (by some otherwise very difficult people) is a testament to Richard Holbrooke's personality, Milošević's political need for a settlement, and PowerScene system's capabilities for visual modeling. PowerScene had demonstrated the value of MP&RS in both combat and peace support operations, and would again in the former Socialist Federal Republic of Yugoslavia in the near future. The U.S. 1st Armored Division used the system in 1996 to familiarize its troops with the Bosnian terrain and to analyze the weaknesses in proposed defensive positions.[18] PowerScene was then used again extensively for mission planning in the 1999 Kosovo campaign. During one media briefing at the Pentagon, Major General Charles Wald of the USAF actually set up PowerScene imagery and F-15E gun camera video of the same target on opposing monitors to demonstrate the system's fidelity and the accuracy of the bombing that could be accomplished with it.[19] In the late 1990s and early 2000s, PowerScene was used to plan military operations over and in Iraq— scenes of Baghdad from street level were developed from satellite imagery to guide the troops who would eventually seize the city.

By 2003, however, PowerScene was declining as a product, having been substantially replaced by Lockheed Martin's Tactical Operational Scene (Topscene) mission rehearsal system, which is now used by all four branches and their special operations forces. A competing product whose roots go back twenty years, Topscene has been installed more than seventy times since 1991, at air stations, headquarters facilities, and aboard aircraft carriers. Cambridge itself attempted to enter a number of related fields—such as video games and automotive navigation—but eventually declared bankruptcy in October 2003—a time of booming military spending—with roughly $1 million in assets and over $6 million in debts. The eclipse of such a lauded product as Power-Scene (and the widely publicized small enterprise that created it) came about rather quickly, and says something rather cautionary about the prospects for similar enterprises in the production of C4ISR) systems.

Evolution of the MP&RS Industry

The aerospace and training simulation market is very large: equipment sales alone amount to about $6 billion annually worldwide. The industry's first and most enduring product has been flight simulators, because the economic case for them is so compelling. Flight hours consume fuel and maintenance for both air forces and airlines; for the commercial operators, training flights also pull aircraft out of revenue service. So, today, the installed base of flight simulators worldwide exceeds 2,100 full-motion units, and tens of thousands of less sophisticated training devices.

While the civil market had been growing slowly, the military market for training, simulation, and mission rehearsal systems increased in volume by almost 20 percent between 2001 and 2004.[20] In general, spending on C4ISR systems drove double-digit revenue increases for many U.S. arms contractors, large and small, in 2003.[21] This growth was driven by some relatively sustainable trends:

Falling cost of computing and display hardware. Rapid advances in computing technology had underpinned technical advances in all areas of C4ISR systems, but MP&RS had gained a particular benefit. By the late 1990s, commercial video games running on laptop computers were providing greater fidelity to players than many of the purpose-built systems of the early 1990s. This meant that basic MP&RS could be carried by every rifle squad and armored vehicle crew for which they were wanted.

Development of sophisticated software engineering tools and libraries. Since MP&RS were increasingly running on generic hardware, the development of more sophisticated software tools meant more widely available simulations and sharply falling marginal costs for additional units. Both trends were positive for military users.

Expansion of special operations forces. Commandos had been intense users of MP&RS for years, as they generally had the funding to afford sophisticated hardware and software, and often had the time to practice their missions before departure. The increased role that these forces had in force structures around the world after the 2001 attacks on the United States bode well for MP&RS firms.

New requirements for simulation in underserved arms of the military. That said, the U.S. armed forces in particular had many segments in which MP&RS had been hardly applied, but where their use could bring great operational benefits. As noted below, truck drivers, firefighters, and maintenance engineers could all benefit from realistic and real-time simulations.

The increasing availability of systems did not mean that their construction was a simple matter. Rather, the development of military-grade MP&RS

requires integration of multiple data-intensive data sources in a computer application whose operation is seamless to the user. These include aerial photographs, digital terrain elevation data, three-dimensional building images, meteorological effects (such as fog and rain), electronic emissions from potential threats, and other operational parameters (such as aerial refueling tracks and navigational aids). Faced with similarly steep development costs, and even lower marginal production costs, the video game industry today is dominated by a handful of firms, such as Sega, Microsoft, and Electronic Arts. It is perhaps unsurprising that the MP&RS industry is also concentrated in the hands of a few firms, including Lockheed Martin (through its Training and Simulation subsidiary), CAE (Canadian Aircraft Electronics, which dominates the full-motion commercial flight simulator business), Flight Safety International, L3 (through its subsidiary Link) and Thales (which entered the business through its acquisition of several simulations firms).

Directions in MP&RS

Ultimately, systems like PowerScene are just three-dimensional viewers, not full mission simulators like those on which airline pilots train. In addition to PowerScene and Topscene, examples of viewing and display systems include Falcon View from the Georgia Tech Research Institute, and Wings and Draw-Land from the U.S. Army Corps Engineers. For more intense training, they must be integrated into full MP&RS like the updated versions of the aforementioned AFMSS and TAMPS. Today, MP&RS are continuing to develop in scale, scope, and fidelity, along several technical and market trajectories:

Faster programming. CAE won a $16 million contract in 2004 to develop a system that quickly gathers all the information needed for mission simulation in sufficient time for relevance to the U.S. Army's 160th Special Operations Aviation Regiment (SOAR). The SOAR had been using TopScene, but this provided no real-time sensor or other threat rehearsal cues.[22]

Real-time support for ground troops. RealSite from Harris Corporation rapidly models ground environments for identifying potential sniper hides and other ground-based threats. The systems was used in preparation for the Iraq campaign; RealSite models of Baghdad were provided to Coalition ground troops before they entered the city. An entire system can cost $100,000 to $5 million, depending on the hardware and modeling involved, but even at the high end of the range, a few units are affordable for a small defense ministry.[23]

Merging mission planning with real-time feeds. Sarnoff Corporation demonstrated in September 2004 a system that projects real-time video feeds from Predator and Shadow unmanned reconnaissance aircraft onto a three-dimensional moving map display using geo-location data.[24] This has considerably expanded the sense of "being there" that helps headquarters staff make decisions based on inputs from their drones.

Firefighting. In 2004, the U.S. Navy installed a simulator aboard the destroyer *Milius* that projects scenes of fire, smoke, and jets of water onto the insides of ship crewmen's firefighting masks. The Harmless Hazards system, developed by the small firm Creative Optics of Manchester, New Hampshire, consists of a grid of magnetic sensors in the overhead television cameras, and the electronic masks. The prototype cost the Navy about $400,000, but this is considerably less expensive than the cost of a fire.[25]

Avoiding ambushes. Lockheed Martin's Virtual Combat Convoy Trainer has been used to train Humvee drivers bound for Iraq. Using commercially available maps and software, the system teaches drivers how to think about choke points and roadside bombs, which is clearly a pressing requirement.[26]

Distributed training. DARPA's virtual schoolhouse project aims to link differing types of users in joint, distributed, and interlinked training exercises. A forward observer at, say, Fort Hood in Texas could call in simulated air strikes via his laptop computer by communicating with a B-52 crew at Barksdale Air Force Base in Louisiana, or even on board an aircraft carrier in the western Pacific, during a break in flight operations.[27] In the long run, widespread adoption of this type of system could even have an impact on base realignments, as troops that once needed to train together might find that they could do so from different parts of the world.

Supporting these technological advances have been not just faster computers, but better geospatial data. In February 2000, the crew of the space shuttle *Endeavor* mapped the entire land area of the earth between 60° north and 54° south latitude using an imaging radar of (hitherto unseen) meter-level accuracy. The results of this Shuttle Radar Tomography Mission (SRTM), a partnership between NASA and the National Imagery and Mapping Agency (NIMA), were subsequently used for the production of extremely detailed digital terrain elevation models. NASA's original idea was to use the data to study floods, erosion, landslides, earthquakes and other geological and meteorological phenomena. NIMA, on the other hand, wanted to use it for planning air strikes, on demand, anywhere in the world.[28]

Small Firms and Big Software

MP&RS long attracted less attention than hardware systems: the products are less tangible, they do not directly kill anyone or break anything, and the spending is quite scattered: most major military acquisition programs have a significant modeling and simulation component built into their costs.[29] Like other C4ISR industry segments, however, MP&RS is a big business, but in several ways, a business more suited *to* big business. Reviewing the criteria from the introductory chapter, we find several reasons for this:

> *R&D-intensive innovation.* The MP&RS industry is driven by the wrong kind of innovation for a small firm to succeed. Flight simulator and mission rehearsal system design is relatively post-paradigmatic: competition today revolves around the use of faster processors and computing cards to display sharper, more complex, more dynamic images for greater fidelity in training. While technology is advancing rapidly, it is doing so in a predictable fashion involving the successive integration of succeeding generations of technology in the input factors. Clever solutions making use of existing technology produce mostly niche products aimed at cost-constrained customers. This is a way to make a living, but not a headline-grabbing enterprise.

> *Skill and capital-intensive production.* Skill is extremely important for the programmers and information architects, but capital is important for undertaking large projects speculatively. This issue of large capital outlays is enough on its own to doom the involvement of small firms, which can only afford so many unsuccessful products in a technologically competitive industry.[30]

> *Widely variable rates of learning.* Software industries combine aspects of both very slow and very rapid learning. During development, programmers have some opportunity to learn from prototype code segments and partial compilations of functionality, but this does not replicate the experience of testing sequential units in serial production of hardware. Once the product has launched, however, marginal production costs are extremely low: essentially those of stamping another compact disk or pressing a few buttons to mail the application across the Internet to a waiting customer. This again is unlikely to provide an advantage to smaller enterprises, even in the absence of the huge-scale economies that appear in business software development.

This is not to say that large enterprises that create software cannot also be inefficient. Complex military software development projects area said to require three important management processes for success: an evolutionary environment, a disciplined development processes, and analysis of meaningful metrics. The evolutionary environment—one in which developers are allowed to pursue small but meaningful improvements—is particularly important. Programs that did not follow this approach included three of those with the most noticeably bad cost and schedule performance in recent years: those of the rather expensive F-22 Raptor stealth fighter, our aforementioned Space-Based Infrared System, and the now-cancelled Comanche attack helicopter.[31]

The system of systems, after all, can be partly defined as one that is constantly, but evolutionarily, changing.[32] Some of the enterprises building it need to be large in scope and scale, and attuned to the right strategy for getting there. Just as Cambridge Research Associates was sinking, the importance of scale in the military IT business was becoming apparent in the fortunes of two other, larger companies: CACI and American Management Systems (AMS).

The CACI-AMS Saga

As John Hillen, then senior vice president and director of the defense and intelligence group at AMS put it, the 2003 campaign in Iraq featured "a lot of new technologies, but in and of themselves, none of them was particularly dramatic. It was the stitching together of all of these things in real time and their integration into the operation . . . that allowed a very small force to essentially take down an entire country."[33] This sort of seamlessness was a big change from 1991, when air commanders had to print out and ship around by air several hundred pages of the daily Air Tasking Order every night. Up until the 1990s, many of the initial developments in information technologies (IT) were funded by the military—the Internet is the biggest and most classic example). Today, however, many of the advances originate in civil developments, not military ones—GPS is the leading example of this dynamic. Alternatively, as Peter Huber notes in the epigraph to chapter 4, the technologies are incubated by commercial firms before they find direct military application. The military IT sector (particularly its C4 component) is thus the sector most exposed to commercial competition.[34]

AMS was founded in 1970 by a group of former Pentagon "whiz kids"— management analysts who had worked in the Office of the Secretary of Defense under Robert Strange McNamara. The company had pioneered the installation of databases and database applications at federal agencies and departments but faced a serious strategic problem in choosing how to tackle the largest military and civil government development projects. AMS had to decide whether to team with other, often larger firms, or to submit its own bids. The company had a history of tackling much smaller projects—most of its contracts were less than two-year deals for less than $30 million—and garnering the opportunity to tackle a large contract is difficult if one lacks a track record of experience with them.[35] Accepting the widespread assumption that the government IT services industry would eventually be split between small, specialized firms and larger firms tackling the largest development contracts, AMS chose initially to attempt organic growth. This strategy did not quite work, but mostly due to several large disputes with clients that predated the September 2001 watershed, and whose adverse marketing impact sharply limited its growth potential. Thus, while most of the rest of the industry was hiring furiously throughout 2002, AMS was laying off 1,000 staff.

The alternative was acquisition. The layoffs, however, depressed AMS's stock price, which in turn limited the firm's ability to buy size by offering its shares as currency in merger deals.[36] The firm thus managed to buy only one firm in two years: R.M. Vredenburg & Co., a small and well-regarded firm that mostly undertook secretive projects for intelligence agencies.[37] Failing that, the best answer for the shareholders was acquisition, but in the other direction. Eventually, AMS CEO Alfred Mockett conceived a plan with Serge Godin, CEO of CGI, a leading Canadian IT services provider, to split the company in two, and sell its commercial and civil government businesses to CGI. The Montreal firm had been willing to buy the entire company, but thought that the Pentagon might balk at having (horrors) Canadians in charge of classified programs. The pair's bankers thought that breaking AMS up actually enhanced its overall value, since the two sides of its business did not interact very much.[38] To acquire the military and intelligence lines, they found CACI.

Founded in 1962 as California Analysis Center, Inc., CACI was a surging presence in government IT services in the 1990s. Facetiously called "Captain and Colonels Incorporated," CACI had experience a similar but more successful growth curve under longtime chairman, CEO, and president Jack London, a former naval officer who saw similar potential in the military market in particular. Finding itself after September 2001 "in all the right niches,"[39] CACI saw the opportunity to capitalize on its appreciating share price with a series of acquisitions that would provide the degree of vertical integration and scope that the company would need to challenge the largest firms in the sector—Lockheed Martin and Northrop Grumman, which also happened to be two of the five largest hardware contractors—and to "stand up with an Accenture or a BearingPoint or . . . Booz Allen"—firms with large commercial practices.[40] CACI also very much wanted to expand its work with the intelligence agencies, an area of particular competence for Booz.[41] CACI's largest previous acquisition, in October 2003, had been of C-Cubed of, a $49 million, 400-person firm in Springfield, Virginia, that concentrated on specialized C4ISR services. In April 2003, it had acquired Premier Technology Group of Fairfax, a $43 million, 360-person firm that focused on intelligence analysis.

In February 2004, CACI bought AMS's military businesses for $415 million, and CGI bought the rest for $443 million. The acquisition increased CACI's work force by 1,700 to 9,000, and pushed its revenues over $1 billion, into a league with firms like SRA and Anteon. As its largest acquisition by far to date, AMS would present some integration difficulties for CACI, and might lead to, some financial analysts thought, a point at which CACI's double-digit annual growth rates would drop.[42] The construction of larger and larger IT systems, after all, indicates how growth in the industry is bounded and self-limiting. As a *Wall Street Journal* editorialist put it in 2004, "in any corporation of any size, there's a department that has grown like a gopher colony on fertility drugs: the 'IT' department."[43] Adoption of complicated IT systems slows progress toward succeeding generations as technologies and systems are absorbed and mastered by staff.

Alliances Are Best between Dissimilar Firms

Small software firms face troubles in whatever sector they operate, as the capital requirements are quite high for large systems and the learning rates are exactly wrong for small enterprises. Moreover, scope economies among related products are considerable for buyers. In many cases, small firms can "survive . . . by exploiting niche markets or by subsisting on maintenance contracts and upgrades" as "an installed customer base will keep these industry orphans alive while offering little opportunity for growth."[44] This is not an exciting vision of the future for them, but some small software firms have created clever and dramatic products. These face the additional problem that alliances are not always appropriate in the military IT field. Reviewing our criteria from chapter 2, we find considerable cause for concern:

> *Relatively stable processes and goals.* As suggested above, MP&RS is one C4ISR field in which processes are not rapidly shifting. User requirements may be expansive, but they are relatively well-defined and can be attained over time with faster processors and better data. Other C4ISR fields are founded on less certainty with respect to the exact methods that will prove optimal and may present better opportunity for alliances (see below).

> *Relatively leaky knowledge.* Alliances among software firms may not work as well as alliances between software and hardware firms—here, the issue is the leakiness of knowledge. C4ISR integration is not a simple question of modularity versus integration. C4 systems in particular are relatively modular with respect to the platforms that carry them, but ISR systems require somewhat greater integration. In these cases, the suitability of an alliance may depend on a wide range of factors.

> *Considerable potential for a shakedown.* Again, among software firms, expropriability is a slight problem, if at all, once standards for interfaces are established. However, expropriability is a bit higher in alliances between software

and hardware firms, where software gets relatively embedded in proprietary hardware products.

Cambridge did attempt an alliance with Silicon Graphics shortly after Power-Scene's performance at Dayton. Cambridge sought the imprimatur of (what was at the time) a marketing powerhouse, and Silicon Graphics "loved how PowerScene made their large systems perform."[45] A comparable example exists today in the alliance that Boeing announced with IBM in September 2004. IBM agreed to supply Boeing with preferred access to high-speed communications chip designs, off-the-shelf computer hardware, and software in return for a ten-year partnership in pursuing "network-centric" development contracts.[46] IBM's experience with commercial middleware would be a great asset to IBM in the rapidly developing area of C4ISR, but neither firm has a fundamental interest in the other's field. Boeing resolutely denied interest in becoming a vertically integrated firm with substantial IT capabilities, and IBM was not about to try to build aircraft and heavy weaponry.

This gets to the question of the suitability of acquisition versus alliance building. IBM itself knows a great deal about the relative merits of alliance versus acquisitions, and procurement versus vertical integration. Indeed, some of the most successful high-technology firms—Cisco, Corning, Intel, IBM, and Microsoft—treat acquisition activity as a corporate competency and on-going activity. While they occasionally pursue large acquisitions—such as IBM's incorporation of Lotus—they are constantly absorbing smaller firms with critical technologies for their portfolios of complicated commercial products, just as CACI did before taking in AMS.[47] However, the question is not merely whether an acquisition buys a useful technology, but whether that technology is more useful to the buyer than the seller. In that event, value flows from the acquisition—otherwise, it can be merely shuffled between firms. Individual industries often move through phases in which one approach—integration or modularization—to the problem dominates.[48] Thus, the importance of recombinative technologies in particular sectors of the defense industry may change over time, so the company may find itself divesting an asset as an easily outsourced function that it acquired years before as a critical capability. In the end, none of these strategies—ally, buy, or get bought—worked for Cambridge. Silicon Graphics had its own problems, Cambridge lacked the resources and sustainable advantage needed to find the financing for acquisitions, and Cambridge's technology was eventually surpassed by the legions of programmers and analysts at Lockheed Martin, who would find little reason to pay for its assets.

For customers, encouraging firms to get their business strategies right is important business in itself, because the other side has some potential to catch up. C4ISR systems drive modern, high-precision warfare, but to some extent, the entry fee into this precision revolution has been cut by the wide availability of electronics and software. As Martin Libicki of the U.S. National Defense University has written,

> Imagine what a sophisticated middle income country could do with a few thousand French and/or Russian precision guided munitions; a few hundred unmanned aerial vehicles (from any of thirty countries); digital video cameras; personal computers; cellular switches phones and pagers; GPS and pseudolite receivers; pocket radars; and night vision goggles; plus archived PowerScene maps combining purchased space imagery and topography, all integrated by a few hundred U.S.-trained engineers—a Radio Shack System-of-Systems.[49]

Drop Your Purse
Force Protection and Blast-Resistant Vehicles

There are a lot of crusades I've been on, and they're all about saving lives. This is a good one.

MIKE ALDRICH, *vice president of marketing, Force Protection*

Introduction

Except to his friends, family, and comrades, Private First Class Alva Gaylord's death might be considered unremarkable: seventeen other U.S. soldiers and marines died that week in Iraq, and more died later from injuries sustained. What made PFC Gaylord's death particularly notable to planners in the Pentagon and the defense industry was his vehicle: he was riding at the time in an RG-31 mine-resistant vehicle (MRV) built by BAE Systems' OMC division in South Africa. The RG-31 crew was part of Captain Jeff Hyde's Company C of the 110th Engineer Battalion of the Missouri National Guard; the team was on a long route-proving drive up and down roads around Baghdad looking for roadside bombs toward to which to vector the battalion's explosive ordnance disposal (EOD) teams. The 110th Engineers had a large area to cover. C Company was based in Scania, a major convoy-refuelling center off Highway One about 100 miles south of Baghdad, and covered the roads all the way north to the capital. The battalion headquarters and B Company were based there, while A Company (attached to the battalion from the 164th North Dakota Engineers) was in Balad, about fifty miles north of Baghdad.[1]

PFC Gaylord wasn't even the first to die in an RG-31: Staff Sergeant Gavin Reinke and Specialist Bryan Quinton of the U.S. Army's 5th Engineer Battalion had died the day before in Baghdad in their RG-31 while passing by another bomb.[2] These deaths were, however, the first three in any MRV in Iraq, and PFC Gaylord wasn't, strictly speaking, in the vehicle at the time. He was riding exposed through the top, behind the ring-mounted machine gun; he died from the shrapnel that the bomb threw in his direction. The vehicle itself "suffered only minor damage in the form of a few chips in the ballistic glass."[3] Had PFC Gaylord been behind the ballistic glass, manning a remote-controlled machine gun by a joystick, he would not have even been seriously hurt. As such, the RG-31 had a very low fatality rate in Iraq, protecting its riders rather

```
NEWS RELEASES from the United States Department of Defense

No. 410-06 IMMEDIATE RELEASE
May 08, 2006 Media Contact: Army Public Affairs - (703) 692-2000
Public/Industry(703)428-0711

DoD Identifies Army Casualty
              The Department of Defense announced today the death of a soldier
who was supporting Operation Iraqi Freedom.

              Pvt. Alva L. Gaylord, 25, of Carrollton, Mo., died of injuries
sustained in Qasr Ar Riyy, Iraq on May 5, when an improvised explosive device
detonated near his RG-31 Mine Protected Vehicle during a combat clearing
operation.  Gaylord was assigned to the Army National Guard's 110th Engineer
Battalion, Kansas City, Mo.

              For further information related to this release, contact Army
Public Affairs at (703) 692-2000.
```

Casualty report on Private First Class Alva Gaylord, Missouri National Guard

better, on a statistical basis, than even Abrams tanks.[4] What is even more remarkable was the performance of two other types of vehicles—the Cougar and Buffalo MRVs—made by a small company located outside Charleston, South Carolina: the aptly named Force Protection Industries. Until December 2006, no Force Protection vehicle had allowed a single fatality.

Early Development

Mine-resistant vehicles draw their origins from the counterinsurgency campaigns of South Africa and Rhodesia in the 1960 and 1970s. The greatest threat to the governments' forces came from landmines, both manufactured and improvised. The weapons were relatively simple compared to the most sophisticated designs sometimes seen recently in Iraq. Most generally had pressure switches, so the explosions were most often under a wheel of a vehicle passing along a rural road. At the time, the primary rural patrol vehicle of the South African National Police was the Ford F-250 pickup truck. In Rhodesia, the police and army were not markedly better equipped. While the F-250 is a rather heavyweight vehicle for the weekend run to Home Depot, a blast under a wheel by an antitank mine could easily throw the truck into the air and reduce it to a heap of scrap metal. The effects on the passengers and driver were even less appealing.

For the South Africans, the most dangerous part of the territory under their control was the Caprivi Strip, the northeastern-most portion of Southwest Africa, or what is today Namibia. The distance by road from Cape Town to Katima Mulilo, the main town in the Caprivi (and on the Zambezi River) is 1,400 miles. As a result, the South African and Rhodesian armies rarely used

FATALITIES IN MINE-PROTECTED VEHICLES ASSIGNED TO U.S. FORCES IN IRAQ THROUGH JULY 2006

Rank	Name	Date	Unit	Location	Cause
			In BAE Systems's RG-31s		
Staff Sergeant	O. Flores	8 July 2006	54th Engineer Battalion	Ar Ramadi	Bomb
Sergeant	A. Floyd	8 July 2006	54th Engineer Battalion	Ar Ramadi	Bomb
Specialist	T. Linden	8 July 2006	54th Engineer Battalion	Ar Ramadi	Bomb
Specialist	J. Micks	8 July 2006	54th Engineer Battalion	Ar Ramadi	Bomb
Specialist	M. Hernianson	23 May 2006	164th North Dakota Engineer Battalion	Al Abayachi	Complex
Private First Class	A. Gaylord	5 May 2006	110th Missouri Engineer Battalion	Qasr Ar Riyy	Bomb
Staff Sergeant	G. Reinke	4 May 2006	5th Engineer Battalion	Baghdad	Bomb
Specialist	B. Quinton	4 May 2006	5th Engineer Battalion	Baghdad	Bomb
			In Force Protection's Cougars and Buffalo		
			None		

tracked vehicles—the distances were just too great.[5] As a result, the South African arms industry came to specialize in wheeled military vehicles; this heritage has continued today in (among other vehicles) the well-regarded Rooikat armored car and the G6 howitzer vehicle from Denel. This fight against bomb-planting guerrillas at great overland distances induced what can be described, as Michael Porter of the Harvard Business School has put it, as a selective factor disadvantage: faced with the serious need to build blast-resistant, wheeled military vehicles, the South Africans came to build very good ones.[6] This is just as well, of course, since tracked vehicles generally ride too close to the ground to avoid blast energy (see below), and thus are at a natural technical disadvantage to wheeled ones.[7]

In South Africa, the development work was led by the government's Central Scientific Industrial Research organization (CSIR). The first approach—a vehicle called the Hyena—simply modified F-250 pickups by installing a wedge-shaped driver and passenger compartment high above the bed.[8] Armor consisted of just 6 mm of mild steel and 50 mm of laminated glass, but formed into a sharp wedge and mounted quite high for adequate protection. The geometry is very important in blast-resistance. Vehicles must be built with enough ground clearance to vent exhaust gasses to the sides, but building the hull underneath into a V-shape helps all the more.[9]

In Rhodesia, a similar effort drastically reduced casualties for the government's troops. In the first 99 mine explosions after the introduction of blast-resistant patrol vehicles, the Army and police suffered only one fatality among 407 passengers and crew. Previous experience for Rhodesian troops would have indicated about 40 dead. This sort of success was duplicated many times over by their minority-rule colleagues to the south—a collection of South African vehicle manufacturers eventually came to build some 19,000 mine-resistant vehicles in a series of types with names like Nyala, Hippo, and Casspir.[10] Eventually, the effort was reinforced by an exodus of relative experts from Rhodesia as well. After the fall of Ian Smith's government to the Zimbabwe African National Union (ZANU) under Robert Mugabe in the elections of 1980, many officers in the former Rhodesian Army migrated south looking for continued military employment. One was Colonel Garth Barrett of the Rhodesian Special Air Service. After a subsequent career in the South African Army, he retired to establish a mine-protected vehicles company with technology licensed from the CSIR. In the 1990s, with the encouragement of the post-apartheid South African government, the colonel began undertaking humanitarian de-mining missions in places like Mozambique, the site of a counterinsurgency campaign by South African and Portuguese troops in the 1970s and 1980s.

At roughly the same time, United States forces were being drawn into conflict in disintegrating Yugoslavia, where large stocks of mines, which originally were intended to stop Soviet tanks—had found their way into fields and roads at the hands of the various warring forces. After NATO's relatively benevolent occupation of Bosnia in early 1996, U.S. troops found themselves dodging these explosives while trying to patrol the country. After the deaths of several soldiers in ostensibly armored Humvees that year, a group from the Defense Advanced Research Projects Agency (DARPA) came to Mozambique to observe a MRV that Colonel Barrett had named the "Lion" sweeping a field with its steel wheels. Each time the vehicle struck mine, a wheel would blow off. Afterwards, it was quickly replaced, and the crew drove on.

The Expanding Threat, and the Evolving Design

This approach, which has been deemed "sacrificing auto parts to save body parts," is a major thrust of, and innovation in, mine-resistant vehicle design.[11] It is also counterintuitive to many combat vehicle designers, who had long built their troop carriers to hold together, not fly apart. Counterintuitive can be good in counterinsurgency, since those waging attrition warfare by planting mines and roadside bombs generally did not follow the same tactics as those leading tank battalions. So, recognizing an important capability that was missing from the defense industry in the United States, DARPA solicited Colonel Barrett's move there, and his subsequent acceptance of citizenship. Before the end of 1996, Barrett and a small team of marketers and engineers had founded Technical Solutions Group (TSG), a firm established for the express purpose of building mine-resistant vehicles. First domiciled in San Diego, TSG quickly moved to Charleston, which had almost as attractive weather, but considerably lower labor costs, cheaper commercial rents, and shorter flight times to Washington, D.C.

The first six years of the company's existence, however, were rough. Marketing conditions were not ideal. While DARPA could provide grease the bureaucratic wheels and sponsor research, it could not buy production vehicles. The Army was supposed to do that, but the late 1990s were a time of both rapid change in doctrinal thinking and stalled procurement. As a result, lots of people were talking, but no one was buying. The talk was of lighter, more deployable forces—particularly after the Task Force Hawk debacle in the Kosovo campaign of 1999, in which the Army tried to deploy an armored brigade to the remote mountains of Albania to protect a helicopter squadron from what turned out to be a fairly unimpressive threat. So, augmenting up-armored Humvees with vehicles more than twice their weight evoked little interest, particularly because the Army was institutionally averse to counterinsurgency. Lighter, however, is not always better.[12] Any vehicle hoping to withstand a mine blast must be relatively heavy: a 20-kilogram explosive detonating under a wheel can throw a pickup truck 25 feet, and does not much less to a Humvee. Even if the occupants are strapped in, the acceleration alone is enough to kill them by breaking their necks.

Of course, weight is not everything—the design of Barrett's MRVs was very clever. As U.S. forces were coming to deal with mines, car bombs, and roadside bombs in Afghanistan and Iraq, Barrett recruited into the company Dr. Vernon Joynt, who had earlier run the entire blast-resistant vehicle research program at the CSIR. While Joynt brought two decades of design experience and *fingerspitzengefühl* to the cause, the basic geometric principles are easy to articulate. In addition to the V-shaped hull, the wheels are kept at the corners of the vehicles, which allows a blast underneath to move to the side. Wheels placed underneath the vehicle—as on a Humvee—tend to channel the blast up into the passengers.[13] As one of his colleagues later put it, reiterating the "sacrificial auto parts" concept,

> when they're hit with a blast, we don't care if the tires get blown off, or an axle gets blown off, or the exterior fittings get damaged. Those are all sacrificial items that can be very quickly repaired and replaced so the vehicle can go back into operation in a very quick turn-around timeframe.[14]

Add-on armor, whether of the self-installed "hillbilly" variety or scientifically designed, is thus of little help against large explosions.

The Vehicles

As often happens in the arms industry, fortunes improved as the problems in the field mounted. In Iraq, the resounding success of the Anglo-American invasion in the spring of 2003 gave way to continuing bloodletting by an insurgency with a seemingly inexhaustible supply of explosives. As shown in figures 2 and 3, by the middle of 2004, roadside bombs had grown into the dominant security problem, and primarily for soft-skinned vehicles.[15]

The Army and the Marines had not fully anticipated the ingenuity of some of the Iraqi insurgents. Landmines had long been a problem in these sorts of wars. Two types of Soviet manufacture had proven particularly nasty. The TMA-3 is not just the most common road mine in the world; it also contains almost no metal, so it is very difficult to detect. If this were not enough, the more recent TMRP-6 sacrifices this stealth for the lethality of a convex steel plate, which on detonation forms a penetrating slug that is very

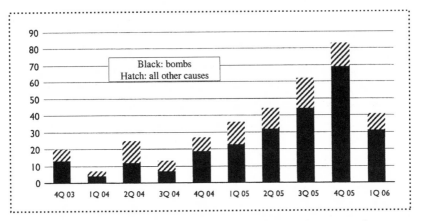

Fatalities in trucks and humvees by cause, U.S. Army forces in Iraq, October 2003–March 2006

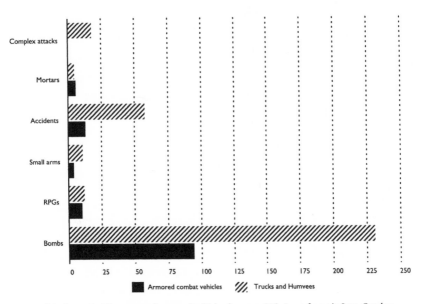

Fatalities in trucks, Humvees, and armored vehicles, by cause, U.S. Army forces in Iraq, October 2003–March 2006

difficult to stop.[16] Mines such as these, though, are largely intended to disable vehicles, which conversely is often enough in a general war, but sometimes less interesting in a guerrilla campaign. To insurgents banking on attrition, dead bodies are needed: in Iraq, roadside bombs could involve wireless telephone triggers strapped to two or three howitzer shells, boosted by a propane tank, and hidden—well above ground level—in the back of a cart or on the back of a donkey. Surviving a blast like that would require a vehicle with more armor and that has roughly three feet of ground clearance below to allow the blast gases to vent.[17]

Sensing both a security need and a business opportunity, the company—which had gone public in 1999 and changed its name to Force Protection Industries—developed and executed an ambitious marketing plan. Two product lines had developed from the work that had started in the mid-1990s and that was about to come to financial fruition. Each was based largely on off-the-shelf automotive components; actually, on commercially procured trucks themselves. The first is the Cougar, an armored troop carrier based on a Mack truck, offered in both four-wheel-four-wheel-drive (4 × 4) and six-wheel-six-wheel-drive (6 × 6) versions. Either version could serve as a blast- and bullet-resistant transport for infantry or engineers, especially explosive ordnance disposal (EOD) teams (bomb squads). EOD versions built for the U.S. Marine Corps and Army featured a ramp for the EOD robot that would generally be carried for inspecting suspicious objects from afar. The second is the Buffalo, which is available for those who find the Cougar insufficiently intimidating. At over forty tons fully loaded, the Buffalo has been described as "too big, too obvious, but nothing better." The Buffalo was designed from the start not as a troop carrier but an EOD specialty vehicle. Rather than carrying robots for standoff work, the Buffalo manually inspected suspected bombs with its spork, an electrohydraulically operated mechanical arm with a steel hand that resembles a gothically curved pitchfork. The spork could excavate, shove, grasp, pick up, and move packages, tires, animal carcasses, and anything else that Iraqi or Afghan insurgents might use to conceal roadside bombs.

Force Protection serially numbered its vehicles, which just happened to befit the individualistic nature of their missions. Some of the individual trucks had impressive service records. As of June 2006, Buffalo #006 had been hit over sixty times by bombs without any fatalities or permanent damage (other than scraped paints and dents). The claw had been blasted off several times, at a cost of $12,000 per shot, which translated to a healthy spare parts business for the company. Buffalo #005 had been almost as thoroughly abused, at one point surviving a blast from a stacked pair of antitank mines, which destroyed the middle axle, but only put the vehicle out of service for ten days.

Common to both vehicles are 800-pound armored windows, 1,500-pound windshields, Foster Miller kevlar spall liners to limit the damage from penetrations by rocket-propelled grenades (RPGs), Look Systems remote cameras, and four point harnesses for keeping the crew and passengers from bouncing around the vehicle during a blast. Force Protection had initially used the same harnesses supplied to NASCAR racing teams, but found that the crash dummies were flying upwards in the blast tests. The company subsequently shifted to aircraft harnesses, which are designed for pitching motions as well. Vehicles built for the U.S. Army vehicles also have overpressure systems for keeping out chemical, biological, nuclear, and radiological contaminants.

Once enough MRVs were in Iraq, the Army began using both types of vehicles in "Trailblazer" or route clearance companies. The individual teams within these companies, which were daily assigned stretches of road to patrol and clear of bombs, were composed of four details.[18] *Sensing* was the job of Meerkats from RSD, yet another South African MRV firm, and a division of car parts marker Dorbyl Automotive Technology. The Meerkat is a spindly 2 × 4 vehicle, with a single driver (and no passengers), that carries both a metal detector and a ground-penetrating radar for locating mines. Like all other South African–designed or –inspired MRVs, the Meerkat is designed to fly apart when blasted, but to protect its driver in the process. *Sweeping* was the role of the Husky, another RSD product. Another single-operator contraption, the Husky is a 4 × 4 vehicle that tows two Duisendpoot mine-exploding trailers—the ground pressure of each is supposedly enough to set off almost all contact-detonated mines. *Interrogation* of suspected bombs (thought too large to just run over) was provided by Buffalos, which (as noted above) were expressly built for the purpose. The Buffalo's success in this mission led to some early exuberance by at least one vehicle crew. After several successful bomb disposals, one Buffalo crew picked up a bomb and moved it in front of the front windshield to snap some pictures. The bomb exploded in front of them, which left the crew shaken up and broke a few bones (and the spork), but otherwise did no permanent damage. *Security* was provided by troops mounted in two Cougars or RG-31s. The latter are lighter MRVs built by BAE Systems' OMC division—also in South Africa. Security is important for passers-by: traffic must be stopped in both directions, a considerable distance away from the site of the operation, and held in place until the issue is satisfactorily resolved—often with a large blast. Security is also important for the bomb disposal teams themselves, who have often been directly targeted by insurgents irritated by their success. At one point in the summer of 2005, guerrillas in Baghdad painted graffiti (in Arabic) "Kill the Claw" on stone walls out of frustration with the prowess of the Buffalo and its spork.[19]

Marketing, and Sales Success

Brilliant products do not always market themselves, particularly when the intended customers are predisposed to avoid buying them. Forcing events help: as noted in chapter 5, interest in militarized aluminum catamarans

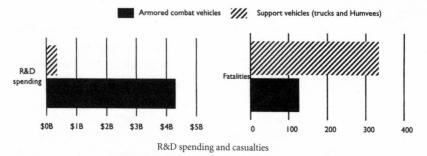

R&D spending and casualties

surged after the Royal Australian Navy's experience on the Dili run in 1999. Forcing events, however, are not always sufficient. This may describe Force Protection's challenge, in that the utility of vehicles like the Cougar and Buffalo had not become fully appreciated by the U.S. Army even three years into the counterinsurgency in Iraq. As late as November 2006, the Army was planning to spend, over the ensuing six years, ten times as much in research and development for armored combat vehicles as for support vehicles such as trucks. Trucks and Humvees, however, accounted for three times as many fatalities in Iraq in the first thirty months of the insurgency.[20]

It was thus shrewd that Colonel Barrett hired as his ninth employee former Army airborne artillery officer and software executive Mike Aldrich as vice president of marketing. Aldrich aimed first at the U.S. Marine Corps, since the Corps is often more agile in its procurement process and has a more unified procurement organization than the Army.[21] The Marines—in matters not directly concerning beach assaults—also tend to have less firmly preconceived notions than the Army about the sorts of wars that their service *should* fight. The focus on the Marines was fortuitous—just as the production run of Buffalos for the Army was coming to an end, and as Marines were falling to roadside bombs in Iraq, Force Protection received a contract from the USMC for 27 Cougar 6 × 6 Hardened Engineer Vehicles (HEVs) in April 2004. Aldrich also invigorated the advertising campaign with frank comparisons of Cougars to lesser vehicles in U.S. service (see below).[22] Trade shows were sprinkled with reversible hand puppets—Cougars on one side, Buffalos on the other—to remind potential buyers or influencers that the company had more than one large land mammal in its lineup.

Like General Atomics Aeronautical and its forerunner (Leading Systems), Force Protection ran through other names—the prosaic Technical Solutions Group and the unlikely Sonic Jet—before settling on a more inspiring name evocative of its mission. During this time, the company also stayed close to its customers: in May 2006, Force Protection had 29 field service representatives

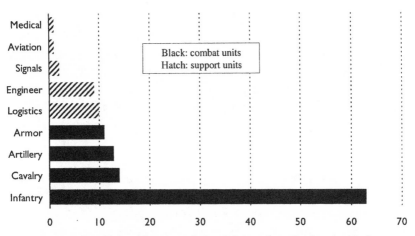

Fatalities in trucks and humvees by unit type in Iraq, U.S. Army forces, October 2003–March 2006

in Iraq providing technical assistance with the maintenance and repair of the vehicles. The cellular manufacturing system that the company initially used to build the Buffalo was helpful in this regard, in that every production worker saw nearly every stage and aspect of the vehicle assembly process.

In May 2005, the company caught its really first big break by winning a program to supply both the Army and the Marines with 4 × 4 Cougars deemed "Joint Explosive ordnance disposal Rapid Reaction Vehicles" (JERRVs). The focus on engineers before the infantry was understandable. As one observer put it, the infantry in Iraq were sometimes "time-wasting targets hanging out on street corners while the EOD teams are busting their butts." This was admittedly an exaggeration, and thus should not be taken to imply that the infantry did not need blast-protected transport. Through early 2006, over half of the bomb-blast fatalities in trucks and Humvees by U.S. Army troops in Iraq were suffered by infantry units. Adding the cavalry, artillery, and armor troops who often supplemented the infantry in its typical roles would drive that figure to over 80 percent.[23]

Protecting infantry closing in armored vehicles on their assault objectives required more than just blast resistance: Cougars (like Buffalos) were built armored all-around to resist most machine gun rounds from 100 meters and to resist overhead fragments from airbursting 155-mm artillery shells. A direct hit from a shell, which would be likely only with a precision-guided round, would do considerable damage to a Cougar or a Buffalo—if not just destroy it outright. Insurgents, however, generally have no artillery and only rarely even heavy machine guns. The Cougar was thus a very affordable candidate for the Iraqi Light Armored Vehicle (ILAV) program, a plan by the U.S. government to equip the Iraqi Army with affordable armored vehicles suitable for chasing insurgents.

In May 2006, BAE Systems won the contract—by bidding an infantry carrier version of the 4 × 4 Cougar *in conjunction with* Force Protection. BAE's own RG-31 armored truck arguably had two difficulties. First, as it was built in South Africa and outfitted in Canada, it would compete against a vehicle that was, as Force Protection's advertisements frequently touted, licensed from South African technology but designed and built in South Carolina. Second, it was arguably a bit lightweight for the duty, as it was more a mine-resistant vehicle than one designed to protect against roadside bombs, whose blasts were generally larger and not always from below.

This is not to say that the Cougar was beyond improvement. By early 2006, several directions of improvement had occurred to the customers, Force Protection, and its competitors. First, a remote-controlled machine gun or grenade launcher turret would greatly improve the Cougar's assault capabilities, bringing it closer in offensive functionality to the Stryker light armored vehicles that the Army had bought from General Dynamics. The situational awareness of a gunner behind his gun was nice, but little could beat protection under armor, and videogame-playing soldiers already knew how to control weapons with joysticks. Worse, several men standing in the exposed turrets of other armored vehicles had been killed in the first few years of the counterinsurgency campaign by bomb blasts that have flipped their vehicles on top of them. Second, neither the Cougar nor the Buffalo was particularly resistant to rocket-propelled grenades (RPGs). The spall liner was useful for limiting the effects of RPGs that had penetrated the armor, but the vehicle itself was not difficult to penetrate. Reactive armor was considered for the Cougar, and even tested by the Israeli firm Rafael, but this would raise the already relatively high center of gravity problem and tend to spray nearby troops and bystanders with shrapnel. And third, the vehicles lacked the electronics that allowed the occupants of other U.S. combat vehicles to find their opponents at long range and in adverse weather conditions. Without infrared, light-enhancing, acoustic, or radar sensors, a fight between U.S. ground troops and the insurgents would be on a closer material footing (at least until U.S. aircraft arrived). Accordingly, in June 2006, GyroCam Systems of Sarasota, Florida, received a $43 million contract for 67 360-degree cameras, plus the associated manuals, installation, deployment blocks, field support, and training.[24]

That said, the Army and the Marines were rather taken by the Cougar as a transporter, at least for constabulary and counterinsurgency missions. Having read the requirements laid out for the Iraqi vehicles, BAE decided that the Cougar was a closer match than its own offering. The story of how the subsequent alliance was formed, which is very relevant to the story, follows.

On Being a Small Company in the Automotive Business

Force Protection's success with its products was achieved while battling many of the classic difficulties that beset small, growing companies. Most notable were the leadership changes; the company ran through five chief executive officers in four years. Colonel Barrett led the company from 1996, but gave up majority control through its initial public offering in 1997. In 2002, Ashford Capital, the new majority owner,[25] decided to replace Barrett; while the colonel was considered a great practitioner, Ashford preferred someone with wider business experience. Software entrepreneur Mike Watts then led the company through 2004, but ran into sudden legal difficulties. Just as the statute of limitations was expiring, Mike and his wife were charged with federal income tax evasion in a matter—not related to Force Protection—regarding their 1997 filing. Since the federal government will generally not do business with a company if one of its officers has been indicted on federal charges, he resigned. Chief operating officer Gale Aguillar then stepped in; he briefly returned to his duties as COO before leaving the company. Scott Ervin, a Texan who served somewhat remotely as the general counsel, then followed briefly as a second interim CEO.

Staff turnover was expected by this point: Colonel Barrett, who had become the chief technology officer after relinquishing the role of CEO, left the company in August 2005 to set up his own rival firm—Protected Vehicles, Inc.—back at the old naval yard. In November 2005, Gordon McGilton was brought in as a permanent boss. McGilton, a veteran of the car parts business at TRW, was resolved to focus on manufacturing operations since the company had repeatedly missed delivery dates of vehicles for the Army and the Marine Corps. At one point, the Cougar assembly line (which converted from a cellular layout after a fashion) featured seven stations—six for assembly and one defined as "rework." While this may have been common in the U.S. automotive industry in the 1970s, successful vehicle manufacturers today generally do not repair brand-new vehicles.

Force Protection company suffered from another problem, rather unusual in the arms industry, though not unknown in the very different information technology business. Force Protection had sold shares at a time in the heady days of the 1990s, a time when "anyone could go public" but when little thought was given to the problems of doing so. The initial public offering (IPO) brought upon the company all of the burden of public ownership with investor scrutiny and public accounting reporting requirements. It also generated all of $600,000 in proceeds, and thus little of the benefits of capital and access to capital that IPOs should bring.[26]

This gets to the question of whether Force Protection's challenges as a small firm outweighed its advantages. Applying the theory from chapter 1, we find conditions, at least temporarily, in Force Protection's favor:

Innovation without R&D intensity. Engineering the Cougar and the Buffalo required a relatively small-scale effort in research and development, at least in the United States. Decades of R&D, licensed from the South African CSIR and stored in the mind of Vernon Joynt, provided an excellent head start. Dr. Joynt was, at least early on, the center of the research effort at Force Protection—he was, as a strategic planner at another defense contractor put it, a man who knew "a lot about explosives . . . *a whole lot* about explosives." Further, while the work was scientifically intensive, the chaotic nature of blast dynamics defied much computational modeling. Practical testing was essential, and Joynt thoroughly enjoyed the work, periodically trekking out to the company's range in western South Carolina to blow up more vehicles. Post mortem hands-on inspection at the range provided the best indication of what new design techniques would work best, since the Army and Marine Corps were providing little combat data back to the contractors on how lighter vehicles were being penetrated.[27] None of this suggested that a large contractor would be able to focus effectively any analytical assets that were not available to a small one, and the practical, low-cost nature of the testing provided a jump to anyone nimble enough to just get to the range and start fiddling with explosives.

Further, the market itself was wide open, and several competing solutions seemed promising, at least to others in the industry. In the summer of 2006, longtime German armored vehicle manufacturer Krauss Maffei Wegmann (KMW) was selling its Dingo mine-protected vehicles to the Bundeswehr, and Israeli newcomer Plasan Sasa was beginning to offer its Sand Cat line, which was based on the chassis of various Ford pickup trucks. Both featured composite armor that was designed to flex in a blast; this technical approach differed greatly from Force Protection's approach with steel capsules derived from South African methods.[28] Blast-resistant vehicles were nothing new in South Africa, but they represented a complete departure in design and oper- ations for North American and European military forces. Further, just what departure was needed in the long run was not obvious. Roadside bombs were the primary killer in Iraq, and RPGs were less a problem than traffic accidents.[29] This was partly because the Soviet-designed and widely-licensed RPG-7 was, though ubiquitous in Iraq, not remarkably powerful. More mod- ern RPGs—that could readily penetrate even heavy tanks—were available from Russian sources, but the Iraqi insurgents simply had not obtained more than a few through the middle of 2006. For example, on 28 September 2003, an M1A1 Abrams tank was hit in an apparent RPG attack that only slightly injured the crew, but started an internal fire that badly damaged the vehicle.[30] If more powerful RPGs became widely available to insurgents, blast protection would not become less useful absolutely, but it could become less a priority relatively. This further suggested that a nimble, get-it-done corporate culture could beat scale or scope economies—at least until requirements became more defined.

Labor-intensive production. Force Protection's factor inputs were heavily skewed toward skilled labor. The large robots common to Ford's and Toyota's factories were completely missing from Force Protection's: parts were installed with chainfalls and moved from place to place with pushcarts. The site in Ladson was a personal affair: welding was accomplished by hand a short walk from the open office spaces in front of the factory, half of whose open desks were staffed by engineers and logisticians. The research staff sat in another building immediately to the rear and in which prototype vehicles were assembled by hand. The only personal offices in the entire company were held by CEO Gordon McGilton and (understandably) his vice president for human resources. The close-knit nature of the operation arguably contributed to the company's sudden rise. Competitors BAE Systems, General Dynamics, and Textron arguably had more bureaucratic systems for decision making. These may be more appropriate for complicated problems of capital allocation or systems integration, but they are not so helpful in responding to emergent requirements. At Force Protection, a frank discussion about the latest prototype could be accomplished by strolling to the back of the property, then huddling a team in a conference room a few yards from the desks of the engineers, the logisticians, and the small sales staff. All of these people were expected to be reasonably technically versed in the product—and physically close to it.

Medium-speed learning. If there was to be an undoing of Force Protection's advantage as a small firm, it was likely to be learning rates. Small companies in the automotive industry generally served two types of niches: small markets for very expensive vehicles, built from the ground up, and slightly larger markets in which existing vehicles could be modified for custom use. Armored combat vehicles could fall into the first category; ambulances and fire engines would typify the latter. Force Protection's success in the market could thus be self-limiting: while the company arguably created the market in the United States, the market could get away from it if the company was too successful. For the blast-resistant vehicles industry, the learning rates would depend strongly on the production runs, and larger production runs would favor larger companies in the long run. Once Force Protection and BAE Systems won the ILAV contract with BAE Systems, the scale and speed of the game changed quickly.

Securing Contracts by Building Alliances

Indeed, while the most recognizable automotive manufacturers are companies closer in size to Toyota, Ford, and Daimler-Chrysler, specialty vehicles are frequently built by smaller manufacturers. Thriving as a small automotive firm often involves striking up strong supplier relationships with larger firms, as these companies have the economies of scale necessary to build basic vehicles that can then be cost-effectively modified by the specialists. Plasan Sasa had chosen to do this with its Sand Cat line: by basing the vehicle on a Ford truck platform, the company ensured that its customers could obtain all other than specialty spare parts wherever a Ford dealer could be found. Blast-resistant vehicles were best designed ab initio to be blast-resistant, but when produced in small quantities, the basic automotive components needed to be derived from vehicles in commercial production simply to keep the costs reasonable.

The use of commercially available parts also provided an exemption from the domestic sourcing requirements of the U.S. federal statute colloquially called the Berry Amendment.[31] Force Protection's basic vehicle was essentially rebuilt from a new-but-torn-down Mack or Peterbilt truck. A wide range of automotive suppliers—and largely not defense contractors—provided the basic components. Axles came from Marmon-Herrington, Mack, Axletech, and Fabco; transmissions were from Allison; tires came from Michelin; and run-flats were from Hutchinson. The engines were commercial diesels from Caterpillar and Mack. The entire thing was painted by Associated Containers of Goose Creek, South Carolina, which also built radioactive waste vessels for the U.S. Department of Energy.

Prior to the ILAV contract, Force Protection's single most important relationship was with Spartan Chassis, the largest unit of Spartan Motors (NYSE: SPAR), a $340 million automotive firm based in Charlotte, Michigan. The teaming arrangement helped convince the USMC that Force Protection could meet its production expectations. As Spartan was producing 6,000 chassis annually, the company had little trouble convincing the Marines' procurement officials that its part of the production workflow would stay on schedule. This reinforcement was very valuable after Force Protection had earlier spent eighteen months delivering the vehicles expected in the six-month contract. Reliability had initially been an issue as well. As the Marines' Buffalo program manager put it loudly to Mike Aldrich, "you at Force Protection have the best mine-protected vehicle in the world, but there are no landmines in my motor pool." Spartan's reputation for automotive quality helped in that regard as well.[32]

SPECIFICATIONS OF FORCE PROTECTION'S THREE VEHICLE TYPES

Feature	Cougar 4 × 4	Cougar 6 × 6	Buffalo 6 × 6
Engine	Caterpillar C-7 diesel	Caterpillar C-7 diesel	Mack ASET AI-400 diesel
Horsepower	330 @ 2400 rpm	330 @ 2400 rpm	450 @ 1800 rpm
Torque	860 ft-lbs @ 1450 rpm	860 ft-lbs @ 1450 rpm	1450 ft-lbs @ 1200 rpm
Sprint speed	> 70 mph	65 mph	65 mph
Highway range	600 miles	600 miles	382 miles
Transmission	Allison HD automatic	Allison HD automatic	Alison HD-4560 P 5-speed automatic
Axles	Marmon-Herrington	Marmon-Herrington	Fabco (front), Mack (rear two)
Tires	Michelin XZL 395/65 R20	Michelin XZL 395/65 R20	Michelin I600 R20 XZL
Run flats	Hutchinson VFI	Hutchinson VFI	Hutchinson VFI
Air conditioning	Dual: 24K and 48K BTUs	Dual: 24K and 48K BTUs	Two units, 44K BTUs
Seating	2 crew, 8 passengers	2 crew, 12 passengers	2 crew 12 passengers
Hatches	2 topside	3 topside	6 topside
Doors	driver, shotgun, and rear	driver, shotgun, and rear	rear only
Height	104 inches	104 inches	117 inches
Width	100 inches	100 inches	97 inches
Length	233 inches	279 inches	323 inches
Curb weight	30,000 Ibs	37,000 Ibs	45,320 Ibs
Payload	4,000 Ibs	12,000 Ibs	38,680 Ibs
Fording depth	39 inches	39 inches	40 inches
Clearance	20 inches under transmission	20 inches under transmission	16 inches under front axle

Spartan eventually wound up assembling entire ILAV chassis to meet the ambitious production goals of the program as Force Protection took on additional work. This rather indicates how the ILAV brought a wholly new dynamic to Force Protection's alliance relations. As it had greater marketing and contract management resources, BAE Systems was chosen by the three firms to serve as prime contractor in the bid. The contract award of $445 million covered 1,050 4 × 4 Cougars to be delivered by May 2007. Meeting that schedule would require splitting final assembly between Force Protection and BAE;

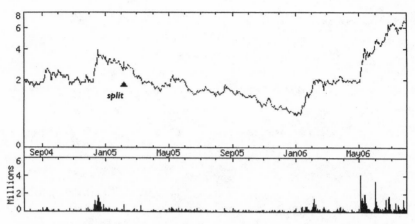

Force Protection's share price chart. The company's share price took investors for a wild ride in 2005 and 2006, slumping with production difficulties and rising quickly with new contract awards.

the latter firm got off to a faster start, assembling the first two vehicle cabs at its factory in York, Pennsylvania, the week after the contract was awarded. York was a particularly flexible facility, with an impressive ability to control unit costs despite the sharp variations of the U.S. Army's funding of its dominant revenue source, the Bradley fighting vehicle upgrade program.

The contract also included two years of contractor logistics support. After that, it was hoped that the Iraqi Army would take responsibility for its own maintenance work, though the record of performance by Arab armies (outside Jordan, at least) suggested that this was ambitious.

Teaming with BAE Systems definitely removed any doubts on the part of the U.S. Army (which was managing the purchase, a gift of the U.S. government, for the Iraqis) that Force Protection could deliver in the ILAV program. The Army had actually been upset with the Marines for buying Cougars from Force Protection when the Buffalo program was starting up: as small and new as Force Protection was, the larger service feared this would overburden the company. After some grumbling at the Pentagon about a rocky first week with the ILAV, Force Protection loudly announced in a late June 2006 press release that the company had indeed met its initial production commitments.

The operational improvements were essential, as more business was headed Force Protection's way. The very next month, British Defence Minister Des Browne announced that the British Army, which had lost eighteen men in lightly armored or unarmored vehicles to roadside bombs in Iraq, would buy 100 Cougars over the ensuing six months. The 6 × 6 Cougars would be outfitted with Bowman radios, British-built electronic jammers, and additional armor. Air conditioning, a standard feature on Cougars not seen on other

British vehicles, would also come in handy in the region. The Cougars were procured in conjunction with 70 more up-armored FV430 tracked armored personnel carriers, in part because the former vehicles would be "less hostile-looking to locals and not as apt to tear up local roads."[33] The order came just weeks after the announcement that the British Army would withdraw from war service its Saxon armored troop carriers some eight years ahead of schedule. The 4 × 4 Saxon was never well-protected, but it was particularly vulnerable to bomb attacks, with blast-traps all over its undercarriage. The Saxons were supposed to be replaced by the first tranche of the Future Rapid Effects System (FRES), a futuristic collection of vehicles conceived as a British analog to the U.S. Army's Future Combat Systems (FCS). In the spring of 2006, however, Lord Drayson, the defense procurement minister, announced that the first tranche of FRES would be met with an off-the-shelf vehicle like Patria's Armored Modular Vehicle (AMV), GIAT's *Vehicule Blindé de Combat d'Infanterie* (VBCI), or General Dynamics' Piranha-IV.

Getting back to the issue of the alliance of Force Protection, Spartan Chassis, and BAE Systems, the interest of the three primary partners extended beyond just the Cougar program, and all three partners would be needed to extend it. Spartan's Road Rescue unit produced some of the best-equipped ambulances in the United States, and several bomb attacks on coalition forces ambulances in Iraq and Afghanistan prompted Force Protection to consider marketing a blast-protected ambulance. Force Protection's technology would be essential here; BAE's role was clear, too. At a meeting in June 2006 (just as the ILAV program was getting started) at Road Rescue's factory in Marion, South Carolina (just a few hours' drive from Charleston), Road Rescue's division president asked whether the team had the production capacity to launch another program, if asked. Mike Aldrich's retort was simple: "Well, with BAE we do." The prototype vehicle was shown at the Association of the United States Army (AUSA) conference, one of the industries largest trade shows, in October.

The alliance was valuable to Spartan as both an introduction and a buffer to defense contracting. The company had little experience in working with the U.S. federal government, and it was quite squeamish about getting involved. As one manager at the company put it at the meeting, "state governments, counties, municipalities, sure. But the federal government? Geez." Despite some recent reforms, the procedural barriers erected by legislation, regulation, and sometimes indifferently educated procurement staff were considerable. On the other hand, the Cougar program was excellent publicity: 80 percent of the questions at Spartan Motors' 2005 annual meeting concerned the alliance and the vehicle. Moreover, if a few shareholders had their own qualms about the company extending itself into defense contracting, the relatively inoffensive nature of the vehicle tended to belay those concerns. Cougars and Buffalos were recognizable as automotive products.

The automotive and commercial nature of the program was helpful to Force Protection, as a small company, in more than one way. Parts were easily obtained: the Mack dealership in Charleston was one of Force Protection's most important suppliers. The reverse was true as well; after stripping down the trucks for conversion into armored troop carriers, Force Protection returned the parts to the dealer—*initially for free*—so that the dealer could resell them as nearly new. The relationships with firms like Mack, Caterpillar, Marmon-Herrington, Allison, and Hutchinson were celebrated with another clever marketing device: NASCAR-style tee-shirts featuring the logos of all the companies running down the sleeves. These concepts worked, but they were probably not the sorts of ideas that would pass the review process at a Lockheed Martin or a Northrop Grumman. Force Protection's relationships were thus often those of a small, scrappy, and relatively new company. The question, then, was whether the company was sufficiently advantaged in those alliances to overcome the problems of scale and inexperience that dogged it. Applying the theory from chapter 1, we find conditions that were actually quite challenging:

> *Considerable change in processes and goals.* Force Protection's market was initially under the radar screen of the larger firms. BAE Systems had bought its way into mine-protected vehicles by buying Alvis-Vickers, which had in turn previously bought OMC of South Africa, which made the RG-31 and similar vehicles. In the United States, however, BAE was occupied for years with the Bradley upgrade program. Primary competitor General Dynamics was busy with Abrams upgrades and Stryker production. Textron was working hard building forty-eight armored security vehicles (ASVs) each month for the Iraqi Army and the U.S. Army's military police units. The ASV was another example of how some of the best customers in the business did not know what they wanted, how they wanted it, or how much of it they would want for how long. The ASV program had been nearly terminated by the U.S. Congress immediately before the war. Similarly, the Buffalo had garnered attention but only enough to sell eleven vehicles. The U.S. Army in particular had been unable to articulate the balance it sought between its desire to be (in some of the parlance floating about) a "force-on-force" army, an "infantry-centric" army, or a constabulary army. In the late 1990s, the Army had sought to be all three simultaneously with the Stryker LAV program, but the shortcomings of the one-vehicle-for-all-roles approach quickly became apparent. Force Protection's difficulties in attracting initial sales were thus understandable, but

they also provided a harbor in which to slowly build the case for its line of vehicles.

Moderately leaky knowledge. The ILAV program was Force Protection's first big opportunity, but it also held the potential for the company's undoing. As noted, BAE Systems had its own line of blast-protected vehicles in the RG-31. While the RG-31 was not as well protected as the Cougar, but this could be remedied with new product development. Just as the Soviet and U.S. space programs each had their own Germans after the Second World War, Force Protection and BAE Systems each had their own South Africans. "Vernon Joynt," one armored vehicle designer observed, "knows a lot about blast, but he's not the only one who knows about blast." The technologies available to both BAE Systems and Force Protection shared a common origin of technology, and many of the attributes were visually observable, such as the shape of the hull, the arrangement of the drive train, and the size and installation of the armored windows. An analogy could drawn with the design of stealth aircraft. Many of the design characteristics were apparent in the shape of the airframe (or the hull), but the details of the materials, joinery, and construction under the skin were not so obvious from a distance. The problem with the ILAV contract was that BAE Systems and Spartan Motors were suddenly sharing final assembly.

Moderate potential for shakedown. By early 2006, Force Protection unquestionably held the leading brand in the United States for blast-protected vehicles. Far more was written in the trade and popular press about the Cougar and the Buffalo than about the RG-31. Force Protection's vigorous marketing campaign led the way, and the U.S. origin of the vehicle and many of its leading components helped. U.S. senator Lindsay Graham of South Carolina, who rather served as the programs' legislative champion, was often keen to note that while the technology and some of the parts came from overseas, the management and the assembly were undertaken in the United States (or, more to the point, his home state). The problem was that while Force Protection had the best vehicle, the best marketing, and the home field advantage, BAE had the contract and the resources. The ILAV contract was the largest blast-protected vehicle award worldwide to date in 2006, but it was not the first, and it would not be the last. For Force Protection, teaming up with BAE Systems removed the next-most capable firm from the competition in the short run, but it reinforced BAE Systems' competitiveness in blast-protected vehicles in the long run. *Ex ante,* in a subsequent competition, Force Protection's bid could be compromised in two regards. First, even without the demonstrably illicit use of any of Force Protection's intellectual property, BAE Systems would gain tacit knowledge of the assembly process and the technology; BAE's staff in Pennsylvania ran a very capable operation and had considerable experience in production process design and vehicle engineering. Second, as the prime contract for the ILAV

program, BAE Systems might, in a future competition for a follow-on program, claim that it had learned from certain limitations of the Cougar, and
chosen to improve its offering with a wholly new design. For a contractor of
BAE's size and experience, this claim could be more than plausible.

All the same, Force Protection chose to forge ahead with not just an alliance
strategy but a strategy of multiple alliances. In November 2006, the Cougar
was declared a "de facto international standard in Iraq" by *Defense Industry
Daily*.[34] Examples of the vehicle were headed for U.S., British, and Iraqi ground
forces fighting there, and the USMC sent Force Protection a $214 million
order for 200 more Cougars and 80 Buffalos. This meant that the vehicles
were selling so briskly that the company had reached the capacity of its facility
in Ladson, and needed more space immediately. So, the company picked
two more partners in rapid succession that month. Armor Holdings agreed to
devote capacity to Cougar production at the former Stewart & Stevenson truck
factory in Sealy, Texas, where the Family of Medium Tactical Vehicles (FMTV)
line of trucks had been built. Armor Holdings faced a rather different set of
challenges: while the company had made a great deal of money producing and
installing armor plating for Humvees and trucks destined for Afghanistan and
Iraq, the future of that market was uncertain. First, the Humvee was widely
criticized as an inadequate vehicle for infantry and cavalry units and was thus
about to be selectively divested from the Army and the Marine Corps' fleets.
Second, bolt-on armor, as noted earlier, was not the right solution to the
problem in the long run. Armor Holdings' interest was thus obvious.

Almost at the same time, General Dynamics Land Systems (GDLS) and
Force Protection formed a joint venture—intuitively named Force Dynamics—to bid on the pending Mine Resistant Ambush Protected Vehicle (MRAP)
program (see below). Force Dynamics' role would not extend beyond this program, in which GD would devote capacity to Force Protection's hot-selling
vehicle at the Joint Systems Manufacturing Center (JSMC) in Lima, Ohio.
This was the former Lima Army Tank Plant at which M1 Abrams had been
produced and were still refurbished, and at which was being built the Expeditionary Fighting Vehicle (EFV), the USMC's replacement for the AAV7
amphibious assault vehicle. While owned by the Army, JSMC Lima was effectively controlled by GDLS. It was also an astute political choice as an assembly
facility, as Army Tank and Automotive Command's rather emotional attach-
ment to the facility meant that any move that secured Lima's workload would
be appreciated. The deal was arranged by Damon Walsh, Force Protection's
vice president for Army programs; as an Army lieutenant colonel, he had been

the commander of the JSMC, so he knew the facility well. GDLS did fit out the RG-31s for the U.S. market under a binding licensing agreement that it had signed with Alvis-Vickers before the latter company was acquired by BAE Systems in 2004. Otherwise, though, the company lacked its own specifically blast-resistant vehicle to market. This suggested that GDLS's involvement had not just financial but strategic value, in that the cooperation could be viewed as another tire-kicking exercise by a larger contractor.

Conclusion: Great Market, Tight Window

Upon the award of the ILAV contract, Force Protection faced two challenges to its impressive run-up as a small defense contractor thriving through alliances. With respect to its size and resources, Force Protection had found one of the most important niches for a company seeking to fulfill emergent military needs otherwise overlooked by the more established contractors. Success, however, expanded the market, potentially laying the groundwork for the larger production volumes and runs that would favor the faster learning curves of larger firms. Force Protection faced a considerable challenge in learning itself how to learn as fast as BAE Systems or General Dynamics could, especially once it had opened aspects of its programs to its most formidable competitors. As larger firms with the strong customer relationships, BAE and GD had the potential to learn so much from its relationship with the smaller armored vehicle builder that it would come in time to no longer need it. General Dynamics arguably had less sophisticated production methods and designs than BAE,[35] but it was still a powerful company in the same market.

For its part, Force Protection was not standing still. By late 2006, the Army and Marine Corps had announced three procurement programs for new blast-resistant vehicles. The first, the Medium Mine-Protected Vehicle (MMPV), would reinforce the Army's engineering units with ultimately 2,500 more vehicles. The Cougar was considered at the time one of the most likely competitors, along with BAE Systems' new 6 × 6 RG-33.

The second, the aforementioned MRAP, proved to be the big prize for Force Protection. Initially planned to bring 600 blast-resistant troop carriers to the rifle units of the USMC, the program plan expanded by April 2007 to encompass some 7,700 vehicles for the Marines, the Army, and even the EOD and construction units of the Navy. The next month, a memorandum from Acting Army Secretary Pete Geren indicated that the program would grow to more than 17,000 vehicles. At less than $500,000 each, the MRAPs would also be very affordable supplements for the Marines to the amphibious EFVs that GDLS was building at Lima for the Marines at roughly $10 million apiece. This was important because, as Mike Aldrich told Defense News, "the production ramp-up being addressed [was] the largest armored vehicle requirement since World War II."[36]

Force Protection was assumed early on to have the inside line on this project: the requirements document specifically mentioned the Cougar as the model against which to bid. Indeed, that April, Force Protection landed the first large-scale production contract for MRAPs—1,000 vehicles, to be delivered in one year, for $481 million. In keeping with the joint venture agreement and to exploit Force Protection's network of alliances, vehicles were scheduled to be assembled in Ladson, South Carolina (at Force Protection); in Lima, Ohio (at General Dynamics), in Sealy, Texas (at Armor Holdings Stewart & Stevenson); and in Charlotte, Michigan (at Spartan Chassis). The aggressive delivery schedule was needed by troops on the ground in Iraq, and it proved the value of the alliances the company had struck over the preceding two years.

Finally, the Joint Light Tactical Vehicle (JLTV) was intended be an all-service replacement for the armored Humvee that would feature a much better level of blast protection than simply steel and ceramic plating. Force Protection had, by the coverage in the trade press, the best-recognized prototype: the Cheetah. Initial briefings from the Army indicated a requirement for over 50,000 vehicles, at a rough price of a quarter million dollars each.

The immediacy of Force Protection's success provides a counterpoint to the long-term view that is often casually espoused by military and business pundits. One fashionable argument about counter-terrorism holds that U.S. defense contractors should try to "build a business model that will last forty years" for the supposed "Long War" against Islamist guerrillas. Rather, some smaller arms makers can profitably think on a shorter time horizon—perhaps three to seven years—since military requirements in the campaigns in Afghanistan, Iraq, and elsewhere are in some flux. Thinking more creatively in the short run could work to the strengths of the smaller firms anyway. Indeed, one lesson for governments in the Force Protection story (or simply the force protection story) is that emergent needs often demand worrying less about domestic sourcing with "full, fair, and open" competition, and instead simply reward innovative companies for taking risks. In doing so, the innovators bring to the military the good ideas that the military has not fully figured out how to ask for.

After all, as Mike Aldrich put it, the roadside bombers had *their* own "international business model, with weapons tested in Sri Lanka, engineered in the Balkans, retested in Indonesia, and finally aimed at troops in Afghanistan." Until the economics of the industry changed, Force Protection would likely continue looking for new ways to counter these schemes with its vehicles, and do so year to year and contract to contract. In addition to working on the essential blast-protective characteristics of the vehicle, Force Protection had put effort into not just protective, but proactive uses of its vehicles. Rather than just ferrying troops on their raids, a Cougar could be configured as a terrorist-hunter-killer machine, with a vapor scanner, thermal imager, ground and wall-penetrating radar, cell phone scanner, remote-controlled laser for destroying bombs from a distance, and a remote-controlled machine gun for dispatching the bomb builders as well.

Indeed, Brigadier Bill Moore, the British Army's director of equipment capability for ground maneuver and the man in charge of the Cougar (of Mastiff, as the British said) purchase, argued that protection from threats of any kind, including roadside bombs, is "thirty percent equipment, sixty percent tactics, techniques and procedures, and ten percent luck."[37] His words echoed a theme that former U.S. Defense Secretary Donald Rumsfeld had reiterated in his tenure in office: military solutions required more than technology, and often could be found simply through intelligent uses of existing technology. Force Protection had done just that with its material solutions, adapting existing technology from South African to threats faced in the Middle East. The troops on the ground were doing their part as well, dealing with the bombs and those planting them with relatively little training. As one company commander in Afghanistan told the Army's official news service, "all our training has been off the cuff."[38] All the same, the troops and the truck builders had together been successful and guardedly optimistic. Through the end of 2006, vehicles built by Force Protection had taken over 1,000 bomb and mine blasts, with apparently only three fatalities, and the loss of only a few trucks from secondary fires.[39] The record stood as its own argument.

The Two Towers
Concluding Advice to Small Firms, Large Firms, and Governments

I often hear the words "future" and "strategic." And when someone comes forward and says, "This is a strategic deal," I immediately say, "No." Often, the word "strategic" can be replaced with "not making money." You have to throw people out of your office if they didn't think things through.[1]

One of the long-standing difficulties with the arms industry is the intermittent nature of realistic feedback on one's efforts. Lately, however, some of the more innovative systems in the Western arsenal have been getting some intensive product testing. The 2003 campaign in Iraq and the insurgency that followed have provided an excellent lesson in how many categories of systems performed and interacted:

Precision-guided weapons (PGWs). These were used in huge numbers. Launched from ground, sea, and air platforms, PGWs provided real-time, adverse-weather lethality with consistency. The variety of PGWs available demonstrated that the trade-offs among competing means of precision engagement had sharpened. Older aircraft had become quite capable when equipped with PGWs, particular those like the Joint Direct Attack Munition (JDAM), which required relatively well-defined and inexpensive integrations with the launch aircraft. Indeed, the war in Iraq was fought with only one fighter or bomber aircraft that was not available in 1991—the F-18E/F Super Hornet, which is to some extent a derivative of its smaller cousin, the F-18C, which fought in the first campaign. No F-22 Raptors or F-35 Joint Strike Fighters were needed to defeat the Ba'athists. Commando forces demonstrated the leverage that they provide the PGM carriers. As the late Art Cebrowski put it, the war demonstrated a "new air-land dynamic: the discovery of a new 'sweet spot' in the relationship between land and air warfare and a tighter integration of the two."[2]

Unmanned aircraft. While still somewhat experimental, their operational effectiveness was confirmed. The huge bandwidth requirements for their video links helped push demand for communications satellite time to an order of magnitude above its past levels. Drones could not, however, solve all reconnaissance and surveillance problems: the Joint Surveillance, Targeting Radar System (J-STARS) aircraft proved absolutely essential for surveillance and targeting during the sandstorm in the second week of the campaign.

Small naval vessels. These were very useful in the confined waters of the northern Persian Gulf. Aircraft carriers were absolutely essential to the victory, since the Saudi Kingdom declined to make its splendid airfields available for the attack, but warships of this size were held back at a safe distance from any naval raiding forces that might threaten them.

Ballistic missile defenses. While improved, these were still having difficulty with even uncoordinated, unsophisticated attacks. Patriot missile batteries in Kuwait and southern Iraq would have benefited from the additional warning time from a satellite system like SBIRS.

Blast-resistant vehicles. Heavily used as well. The exploits of the "horse commandos" in Afghanistan notwithstanding, armor still provided a great deal of survivability and lethality on the ground and demonstrated their long-range mobility in the motor march on Baghdad. The type of armor required, however, evolved with the campaign. After the defeat of the Republican Guard in the field in 2003, Coalition forces faced almost no antitank missiles or cannons. Insurgents deployed rocket grenades and (more dramatically) roadside bombs, but purpose-built vehicles proved more useful than the Bradley and Warrior troop carriers that the U.S. and British Armies had brought initially.[3]

Friendly fire was also still a problem, despite the great increase in information available. The persistence of this issue should be taken as a warning that C4ISR linkages are not created effortlessly but require a great deal of both systems integration and product and process innovation.

Small Firms Should Aim Carefully

Many of the advances in the war—whether the massive re-equipping of older fighter-bombers with newer, autonomous PGWs, the basing of naval commandos on a militarized car ferry, or the use of unmanned reconnaissance aircraft to scout targets for the artillery—demonstrated the utility of the dynamic recombination of existing technologies. Small, dynamic firms often excel at adapting new but off-the-shelf systems to other off-the-shelf systems, and which returns us to the question of where their advantages are to be found in the new arms industry. Potential entrants should consider three questions in trying to determine whether a small firm will possess a relative advantage:

CAN AN INNOVATIVE BUT NOT RESEARCH-INTENSIVE
PRODUCT SUCCEED?

Recombinative technology often means innovation without constant attention to high science, but not every segment of the industry will be equally
given to innovative activity by small and medium-sized enterprises. Ceteris
paribus, given a certain level of small-firm innovative activity in an industry,
the greater the amount of total innovative activity in that industry, the less
the technological environment facilitates entry.[4] Small firms should seek the
less-trodden paths in search of unknown, even initially unknowable opportunities. That is, if one knows "everything there is to know about a [new]
product, it's not going to be a good business. There have to be some major
uncertainties to be resolved. This is the only way to get a product with a major
profit opportunity."[5]

There are those who doubt that this recombinative innovation can be sustained. In the past, major periods of military innovation have often followed
stinging defeats. If al Qaeda's attacks on the United States in September 2001
can be characterized as such, then the mid-term results seem to have been the
removal from power of the Taliban and the Ba'athists from Afghanistan and
Iraq. All the same, while innovative products have been more heavily used, no
broad national research agenda has emerged in the U.S. fight against political
Islamism or resurgent Ba'athism. On the other hand, we do know that serious military innovation occurs when civil authorities get involved in steering
military authorities toward more relevant responses to threats.[6] Innovators in
uniform are often midcareer officers, around long enough to care but not so
long as to have become cynical or entrenched. Senior officers are frequently
not the source until their careers are threatened by reform-minded civilians.[7]
Even if progress has been uneven, this unequivocally began to happen under
former U.S. Defense Secretary Donald Rumsfeld, building on more tentative
steps taken under Secretaries William Perry and William Cohen in the 1990s.
Some changes in senior staff may have signaled throughout the officer corps
in the United States and the United Kingdom that risk taking and inventive
thinking are to be rewarded in the future, though this is hotly debated.[8] Perhaps most importantly, two of the most significant examples of innovative
military technology in the past fifteen years—the Predator and the JDAM—
were produced under very tight budgets. More can indeed be done with less,
and small firms are often at least as good as large ones at doing it.

WILL SKILL BE MORE IMPORTANT THAN CAPITAL?

The question of how many units will be produced is a large part of this consideration. Large production runs argue for mechanized assembly, which can be capital-intensive and thus the province of large firms. General Dynamics and (later) Lockheed Martin made a great deal of money from the F-16 Fighting Falcon program, and Lockheed Martin now stands to earn a great deal again from the analogous F-35 Joint Strike Fighter (JSF) program. Bell Helicopter has had a similar ambition with the Armed Reconnaissance Helicopter (ARH) program, which one manager there described as something "we're hoping to make into the F-16 of the helicopter business."[9] Many programs, however, have built-in constraints that limit the use of large production volumes. Predator aircraft consume so much satellite communications bandwidth that the USAF can only fly a few at a time in any part of the world. Even if the service wanted a thousand of them, it really has no use at this point for more than a few hundred at a time—unless a disruptive technological advance can resolve the problem. In Robert Clifford's sales dream—and it is good to dream—the U.S. Army and Navy could buy one hundred of his fast catamarans. Though the services could conceivably buy a thousand, they most likely will not. Incat and Austal thus need not worry about Ford and Toyota usurping their production system.

The evolution of military forces' procurement practices are at least as important in this regard as the evolution of the dynamics of warfare. If procurement remains centralized and subject to long product cycles—like the triservice, multidecade Joint Strike Fighter program—then small firms should find other work. If, however, experimentation, spiral development, devolved procurement authority, and short product cycles take greater root—as with the Army's and the USMC's parallel pursuit of multiple types of blast-resistant vehicles—then small, entrepreneurial firms will find an environment in which to thrive. Many of the Pentagon's emergent requirements in campaigns from Iraq to Bosnia and back—such as those in proximity fuze jammers, hand-launched reconnaissance drones, and ceramic body armor—have been met by small and medium-sized enterprises.

CAN WE LEARN THROUGH PRODUCTION, BUT NOT LEARN TOO FAST?

The question of how fast the production runs will proceed is a significant part of this consideration. Small firms have succeeded in marketing armored vehicles, but this may not be the best industry for small firms, and the reason for this is akin to the answer to the preceding question. Still, even armored vehicles are being produced in spirals of development in which critical sub-systems will be upgraded far more frequently than the platforms themselves are replaced. LAV-III armored cars around the world are equipped with remote 12.7-mm machine guns, manned 25-mm automatic cannon turrets, wire-guided missiles, 120-mm mortars, and 105-mm assault guns. None of these weapons are made by General Dynamics. Procurements like these require considerable systems engineering and integration work, but they also require well-engineered components and assemblies, which in many cases are the province of smaller specialists.

Fortunately for those considering the plunge into the military-industrial complex, being small in this business is increasingly working out. While some European governments are still trying to figure out how to protect their national champions, others have moved on. The Swedish Defense Material Administration is quite content with the state of its admirable collection of arms businesses, which balances consolidation with an openness to foreign investment. As the head of Swedish military procurement put it a few years ago, "we do have a defense industry in Sweden, but we don't have anything like a Swedish defense industry."[10] In the United States, large-scale consolidation may have run its course, at least for some time. The money is starting to follow: while large firms' share of procurement funding has stayed constant, small and medium-sized firms have been chipping away at their share of research, development, and test and evaluation (RDT&E) funding.[11] This presages greater involvement in the actual production of systems, since contract responsibility is not frequently moved from one firm to another when development shifts to production (the arms industry is not quite the pharmaceutical business).

MARKET SHARE OF THE TOP TEN PROCUREMENT
CONTRACTORS BY PHASE, ADJUSTED FOR
ACQUISITIONS

	1996(%)	2001 (%)	Change (%)
Production	77	77	0
RDT&E, total	64.1	58.7	−5.4
RDT&E, early stage	45.6	37.1	−8.5

This is not, however, the start of an outright revolution. As noted in chapter 2, arms making remains subject to daunting scale and scope economies in many fields. The remaining question for small firms is how to get involved, whether alone or with a partner.

	Advantage to the Small Firm?	Suitable for Alliance?
Space satellites	Possibly	Probably Not
Unmanned aircraft	Yes	Yes
Guided missiles	No	*
Small warships	Yes	Yes
Mission planning and rehearsal	Possibly	Probably Not
Mine-protected vehicles	Possibly	Possibly

Small Firms Should Choose Their Friends Carefully

Many of the best small firms in the arms industry have excellent technology but considerably less marketing and systems integration capability than the established industry heavyweights.[12] As the table above indicates, some sectors of the business are appropriate not just for small firms but for alliances of firms large and small. Small firms that think that they can succeed in the business should consider this option, but beforehand, should ask three questions. Fortunately, some of the factors that suggest a role for small firms also indicate a role for alliances as well:

Are the customer's requirements unsettled?   Innovative, widely adaptable products and clever operational methods are often the way to solve problems that have defied purely technological solutions. They are particularly useful when customer requirements are moving targets. This sort of market may scare off the weak-kneed, but it also presents an opportunity for the clever, fast moving, and organizationally adaptable to make money by banding together for defined activities.

If, for example, a land force suddenly needs a lightly armored, high-speed, fire support vehicle, an armored vehicle manufacturer can team up with an ordnance maker to meet the requirement more quickly than either could alone through a market purchase and subsequent integration of the other's product. Even before becoming parts of BAE Systems, Swedish armored vehicles maker Hägglunds and Swedish ordnance maker Bofors collaborated on the development of a number of weapon systems in this way. If the customer instead needs a lightly armored, high-speed tank destroyer, the armored vehicle manufacturer can shift his alliance to include a guided missile manufacturer, and put his other relationship on ice. A fully integrated firm—one with armored vehicle, ordnance, and guided missiles divisions—may not benefit from the wider scope of internal activities *unless its customers' plans remain relatively stable.*

Is our knowledge reasonably leaky? If the products that military customers need are completely plug-and-play, then one firm should simply sell black boxes to the other in a market transaction. If, on the other hand, customers require tightly integrated products, then design concepts and production knowledge would need to be shared rather fully between the firms, and the risk of appropriation would be great. In that case, a merger of the two firms is called for, as this will align the interests of the managers in both.

In shipbuilding, integration problems are significant, but they generally do not intrude on shipbuilding matters per se. The bending and joining of aluminum has little to do with the installation of modular missile launchers and naval guns. Thus, Incat finds itself in an alliance with Bollinger and not with BAE Systems. In space satellites, the degree of integration required is so great that alliances have significant organizational costs: TRW and Spectrum Astro's efforts to organize development of the SBIRS Low in this way both failed, and both firms wound up acquired by two of the largest firms in the industry. In short, alliances with dissimilar firms are very useful, but only in particular sectors and subject to particular design paradigms. Sometimes, just buying the outfit makes more sense.

Are we each slightly vulnerable to a shakedown? This is a matter of modularity. For an alliance to be the suitable form of organization, integration with the larger system or system of systems must be close, but not too tight. Indeed, this rather describes the ideal alliance as well. If integration is too loose, then the systems are weak complements for one another—one could, again, say this about warships and naval guns. Most naval guns fit on most large warships, and indeed are usually specified by the end customers, so neither shipbuilders nor naval ordnance makers have the upper hand in the transaction.

If, on the other hand, the integration with the larger system is so tight as to preclude its use with other systems, and the firm with the larger system relatively owns the customer relationship, then the larger firm will have the upper hand in price negotiations. Here, the subsystem manufacturer has limited options outside the relationship, but the platform manufacturer may be able to specify another subsystem in his next proposal. Many of the small suppliers that found themselves in this sort of position were acquired by their customers in the vertical integration drive of the late 1990s. This was entirely appropriate, since the smaller firms of this type had limited options outside the embrace of the larger ones anyway.

These three factors are interrelated but not in an obvious way. Increasing uncertainty in military requirements will tend to increase the modularity of the products demanded, which will in turn tend to decrease intellectual appropriability and profit expropriability. Modular subsystems require a lesser degree of openness in specification than tightly integrated ones and are more easily ported from platform to platform. The question for alliance suitability, then, concerns the starting point. If ideas were quite closely guarded, and if profits margins of suppliers were high, then increasing uncertainty about requirements will point toward alliances and away from vertical integration as the efficient form of organization. On the other hand, if ideas were leaky to begin with, and profits hard to maintain, then increasing uncertainty in requirements may not indicate alliances. Any weapons contracting firm considering this problem needs to investigate the current structure of the industry *and* the relative certainty of future customer requirements before reaching a conclusion about how to proceed strategically.

STRATEGIC CHOICES FOR LARGE FIRMS FACED WITH SMALL, CAPABLE COMPETITORS IN THE ARMS INDUSTRY

		Advantage to an Alliance?	
		Yes	No
Advantage to the Small Firm?	Yes	ALLY with the small guys	ACQUIRE their technology
	No	Find another ally, then ATTACK	ACQUIRE the whole company

Strategies for Interaction between Firms Large and Small

At this point, we can put together the analysis of both firm size and alliance suitability to draw some conclusions about corporate strategy. We will treat this from the standpoint of the large firm, but the observations apply to small ones as well: the viewpoint merely must be reversed. Large arms makers should care quite a bit about small companies (and vice versa), not out of any sense of generosity but because they both pose competitive threats and present business opportunities. If systems integration will be an essential skill, and if widespread small-scale experimentation takes an important role in the formulation of future forces,[13] then large firms should think about how to interact with small firms. The best of the small firms tend to excel at finding clever, cost-contained solutions to difficult problems. Within the context of our discussion, large firms have four options with respect to small ones (these are summarized in the figure 8.4). While many other considerations intrude upon portfolio and strategic planning, the relative indications of the factors above, ceteris paribus, are these:

Ally with the small guys. If the advantages in the industry lie with small firms, and the conditions suitable for alliance formation are present, then a small company with outstanding technology can make a great ally.

Acquire the technology. If the advantages in the industry lie with small firms, but the structural conditions for alliance building are not present, then a small firm with the right technology can still be a valuable supplier. Since large organizations may be relatively disadvantaged in this case, a market relationship may be preferred. A long-term supply contract with the correct incentives can provide some of the benefits of vertical integration without the accompanying administrative difficulties and managerial disincentives. Eventually, the company *could* be acquired, if other factors endemic to the arms industry, such as domestic customer access or multidomesticity, indicate the value of acquisition. At that point, the firm can be managed as an independent subsidiary, which in the latter case would make the national government of the smaller firm's domicile rather happy.

Find another ally, then attack. If the small firm criteria do not apply but the alliance conditions are present, then consider finding another ally—perhaps a large firm with similar technology—and competing. In the short run, the small firm may make a useful supplier; in the long run, it may exit the industry as its scale and scope inefficiencies are laid bare. At that point, the remnants of the firm and its technology could be acquired quite cheaply. Alternatively, and as in the next case, the smaller firm can be acquired sooner—and perhaps more Pareto-efficiently—if its management recognizes the predicament.

Acquire the whole company. If the small firm criteria do not apply, and the conditions needed for alliance formation are not present, but the small firm still has excellent capabilities, then those can be subsumed within the larger firm by outright acquisition. The price may be right if the leaders of the small firm can be convinced of the long-term futility of their quest: their small firm is unlikely to thrive in that environment, and no big brother will munificently shelter them under its wing.

At this point, the question becomes one of how to manage the acquisition. As noted above, some of the most successful high-technology companies—Cisco, Corning, Intel, IBM, and Microsoft—treat acquisition as a corporate competency, constantly absorbing small firms with critical technologies for their portfolios of complicated commercial products.[14] Acquisition is thus an ongoing process, and not because all high-technology industries are relentlessly and forever consolidating.[15] Building these capabilities is not simply a matter of agglomerating an industry. As the arms industry has seen a great deal of merger and acquisition activity since the end of the Cold War, not every additional deal will be as attractive as the average one was ten years ago. In the 1990s, the success of Lockheed Martin's and Raytheon's respective acquisition sprees notably stalled as the respective headquarters lost grasp of all their disparate operations were doing—as noted, large and loosely governed entities tend not to perform well economically.[16]

This and the rest of the preceding discussion suggest at least three factors that managers in the arms industry and policy-makers in government should consider in evaluating the suitability of additional acquisitions:

The right balance of knowledge spillover. This is particularly important for the ultimate customer, the government. Despite frequent concern in merger reviews over the potential for vertical restraints, it is quite unlikely that one arms firm would refuse to supply critical technologies for the product of a rival after the latter firm had secured a contract for a major system or platform. The risks of alienating the government, the ultimate and monopsonistic customer, are huge. Governments also have a habit of demanding and enforcing consent decrees for chinese walls to suppress potential vertical restraints. However, any assimilation of a small firm into a larger one tends to restrain the *pace* at which new ideas circulate in the community of engineers and technical managers. As suggested above, the strength of the industry requires that knowledge be leaky but not too leaky. The independent survival of some firms—though not all—particularly as subcontractors, is essential for maintaining the pace of idea generation. Again, this is particularly important in fields of rapidly developing technology (though perhaps not technology that is too research-intensive).

Progress toward stable processes and objectives. Again, relative stability in industries indicates closer integration as an ultimately more efficient form of organization. This is true even if the technology is relatively advanced, so long as the direction of its development poses few questions. The SBIRS Low project, described in chapter 2, provides a clear example. The competing teams were consolidating in areas of work such as the satellite bus and the ground control environment, but two contractors were left to compete for the right to supply the infrared sensor. While other portions of the satellite system present challenging systems integration problems, success is more certain in those matters than with the payload, whose solution may be found in relatively unexplored scientific territory.

Integration at a brisk but manageable pace. If acquisition makes sense, then the remaining question is one of speed. The incorporation of target companies into a large systems integrator is probably best done quickly and decisively—or not at all. The effective management of acquired companies and subcontractors has been cited as one of the reasons that Lockheed Martin beat Boeing for the Joint Strike Fighter project, potentially the largest military acquisition program in history.[17] On the other hand, Northrop Grumman was relatively successful with its initially arms-length management of Newport News Shipbuilding & Drydock; closer integration of the shipyard, including adoption of the new parent company's SAP enterprise software system, was initially put on hold pending the successful integration of other assets. Shipbuilding and unmanned aircraft production (for example) are not too closely related, so Northrop Grumman can afford a go-slow approach. If a merger makes sense in the future but the financing and will are present today, full integration can be deferred until the managerial resources are available to handle it.

Governments Should Care about Small Firms—and Not Sentimentally

This leaves advice for governments—the customers, who should be as interested in industrial strategy as any large corporation should care about supplier management. Small firms do garner attention from industrial strategists: Europeans and Americans have a cultural proclivity for underdogs, and the arms business is no exception. In the United States in particular, a considerable volume of government procurement contracts are set aside each year for all manner of small businesses. The small business lobby, after all, is one of the more noted in Washington. All the same, the Pentagon's interest in small business can extend past the sentimental and the crassly political.

First, consolidation in the arms industry has limited defense ministries' options. After fifteen years of consolidation in the United States, the effects are apparent: there are only so many domestic suppliers in any given sector. Consider space boosters: in the summer of 2003, the USAF stripped Boeing of nearly $1 billion of future satellite launches and awarded the contracts to rival Lockheed Martin. The penalty was decreed after it was discovered that two of Boeing's employees had written Boeing's proposal for the work using some of the 25,000 confidential documents that they had pilfered from their last employer—Lockheed Martin. The incident occurred in 1998, when Lockheed Martin was clearly the dominant competitor for USAF launches, so the initial award rather surprised many observers. The USAF also suspended all further contract awards to Boeing's space division, *but only for 60 to 90 days.* After all, Boeing was just too big to avoid doing business with.[18] In 1990, there were six contractors supplying booster rockets in the United States: Boeing, Lockheed,

General Dynamics, Martin Marietta, McDonnell Douglas, and Rockwell. In 2003, there were only two.[19] If not Boeing, the only domestic choice was Lockheed Martin.

In Europe, this problem has been apparent for decades. From the 1960s through the 1980s, Britain, France, Spain, Italy, and Germany each tried at times to advance the export potential of a small set of large, local firms considered "national champions." Britain, under the liberalizing aegis of Prime Minister (now Dame) Margaret Thatcher, was the first to ditch this philosophy. In December 1984, Sir Peter Levene (now Lord Levene of Portsoken) was appointed chief of defence procurement by the British government. Demanding cost and quality performance, and inflicting financial penalties in the absence of them, Sir Peter managed "to break the cozy relationship between industry and the [Ministry of Defence] that was not healthy for either side." While Mrs. Thatcher admitted to "going to bat for Britain" industrially from time to time, the discipline imposed on British arms makers may have spurred their export successes in the 1990s.[20] Recently, however, the problem may have returned through an overly aggressive assumption of the "strategic" importance of BAE Systems to the Crown. Manfred Bischoff's epigraph provides an alternative definition of that word.

Second, smaller firms often produce the particular kinds of technical advances that military forces need in the current strategic environment. Common reference often considers technology an exogenous factor whose "level" progresses uniformly upward through time. Technological progress, which is quite endogenous to our problem, often occurs in fits and starts in different sectors, and it is subject to dynamic feedback as well. Naturally, the supply of solutions depends on the economic incentives provided to a given supply curve for scientists, engineers, and programmers. More to our purposes, the types of technologies that are developed depend on the structure of the industry that produces them. As described earlier, smaller firms thrive in industries in which progress is evolutionary rather than revolutionary. This is not to say that small firms do not thrive during times of rapid technical progress, but that the progress under which they thrive is continuous, incremental, and recombinative. This sort of technical progress will beget small, dynamic firms, and these sorts of firms will then subsequently produce this sort of progress.

This points to a potentially important role for smaller, entrepreneurial firms. If technological advances are increasingly about dynamic recombination of existing technologies, then smaller firms *may actually help produce* clustered waves of the sort of continuous, incremental, recombinative innovation that can be particularly helpful in times of high uncertainty about future requirements. If military customers are particularly interested in this sort of material progress, then encouraging small firms to get involved in the arms business would be good industrial strategy. Appropriately, the U.S. Department of Defense has announced that it wishes to increase its involvement with second- and third-tier suppliers (who are often smaller firms), and to provide incentives for prime contractors to work with the most innovative among them.[21] Implementation of this intent is another matter. Policy can be promulgated and guidance for program managers published, but overall small firm contracting (other than the existing outright set-asides) may not increase if interest in working with smaller firms is not generated program-by-program.

Still, there are general actions that can be taken at the defense ministry level to increase the chances for small-firm innovation without mandating a particular contract award:

At a minimum, do no harm to small firms. Doing business with some defense ministries is not easy; doing business with the Pentagon can be epically difficult. Small firms have a relatively disadvantage with respect to these barriers because they lack the scale necessary to employ enough contract administrators and lawyers to navigate the military acquisition process. Simplifying the process would encourage more small and commercially oriented firms to do business with the military, but since this advice has been in the literature for decades, there is little reason to say more at this point.

Experiment, experiment, experiment with novel applications and combinations of weapons and platforms. Many more firms can be supported by research projects and short production runs (like that of the Cougar, ASV, and RG-31 armored vehicles) than can be supported by twenty-year procurement programs (such as that of the Future Combat System). As Jacques Gansler, formerly under Secretary of defense for acquisition, technology, and logistics, put it, maintaining a long-run competitive presence in armaments is not about "20,000 people in production; it's 100 really smart engineers."[22] Affecting that change, however, would require a fundamental commitment to ongoing experimentation that has not been seen in the United States outside the Advanced Concept Technology Demonstration (ACTD) program. In actuality, the ACTD program is not about high science or groundbreaking new component technologies. The Predator drone was developed through the ACTD program, but all its major components were composed of relatively off-the-shelf technologies. It was the recombination of these systems in a novel form that produced such a low-cost and useful weapon.

Invite industrialists to help define requirements. The arms industry, to be sure, has been the source of many innovations hatched outside the traditional requirements process. The idea behind the Predator reconnaissance drone and the use of the *Jervis Bay* as a military transport originated with the firms that developed the products. The armed forces in the United States and many other countries have traditionally not asked industrialists and engineers to participate in defining its concepts of operations. This exacerbates the problem posed by the frequent unfamiliarity with the technical state of the art among the operational staffs writing the operational concept documents.[23] End users as well should be involved in the development early on, and this does not mean just asking the staff people writing the requirements.[24] Fortunately, this is changing—modern practice generally expects that software engineers be involved in developing operating concepts for the commercial organizations on whose behalf they work, and much of the functionality of modern weapons is found in their software.[25] This does not mean that military services should surrender responsibility for trade analyses to contractors. It merely means that early involvement in requirements development by small, innovative firms with clever operational concepts can enhance operators understanding of the possible (e.g., the *Jervis Bay*) rather than merely the stipulated (e.g., another tank landing ship).

Stiff-arm requirements creep. The success of the Predator, the aluminum catamaran, and the Cougar transport have depended strongly on the relatively narrow scopes of their missions. This enabled a wide set of users to field the equipment much faster than would otherwise have been expected. Faster fielding creates more opportunities, sooner, for experimentation in the field, and this has an option value all its own. Two retired admirals agree on this matter with respect to the Joint Tactical Radio System (JTRS). The late Art Cebrowski, one of the two directors of the Pentagon's former Office of Force Transformation, and Bill Owens, formerly vice chairman of the Joint Chiefs of Staff, and later CEO of Nortel Networks, have each argued that the U.S. Army and the USMC could have deployed very capable wireless systems based on ruggedized, commercial Internet-protocol communicators to troops in Iraq in time for the campaign in 2003. Instead, the troops had to make do with older radios of lesser performance while the JTRS program continues its all-encompassing development process. Three years after the program launched, no equipment had reached the troops, and users who urgently needed upgrades were needing to secure waivers through Washington for orders that did not flow through the JTRS program office.[26]

The Possibilities of Commercial Integration and Vertical Deintegration

This last recommendation points to one of the more exciting aspects of enhancing the involvement of small firms, particularly in the information technology arena, in military-industrial activities. Properly encouraged, commercially oriented firms (whether as large as Nortel or far smaller) may more quickly port commercial technologies into military products, and on development cycles more akin to wireless telephones than the JTRS. Many casual but well-placed observers have claimed that military production, unlike civilian production, is endemically characterized by low volumes, custom orders, a high degree of dependence on a few customers, and a small number of rivals. High-profile studies have claimed that the collection of industries supplying military materiel, in particular to the U.S. armed forces, were isolated from the rest of the economy, and that the situation was not likely to change.[27] Actually, arms firms, particularly smaller ones, are not entirely isolated from civilian production. At the end of the Cold War, over 80 percent of all plants with Department of Defense contracts integrated military and civilian production in the same facility, and almost half of all production plants in the machining-intensive durable goods sector in the United States had contracts for military goods.[28] The truth is that many segments of industrial economies have these features, but these features do not fully insulate the industries from discontinuous technological change imposed by new entrants with other ideas.[29] Further, there is little new to the idea that strength in military industries has had beneficial externalities toward civilian industries.[30] Separation is particularly noted in aircraft, but not in those segments where commercial transports are adapted for military use, as tankers, signals intelligence, and command and control aircraft.[31] Many component industries, however, particularly in electronics, are already well-integrated, and getting more so.[32]

There is little reason to forestall this, even if one wished that forestalling were possible. That is, in a particular combination of circumstances—

- short-term constraints on skilled labor,
- a reasonable leakiness of product and process knowledge,
- uncertainty in customer requirements but demand for innovative solutions, and
- innovations that produce relatively modular systems but that still require somewhat complex integrations.

The key will not be found in vertical integration, but in well-chosen alliances with small, sophisticated firms. Those circumstances are precisely the ones that are overtaking some sectors of the arms business today. Innovation is changing the nature of the industry today, and in some sectors, is reducing the minimum efficient scale of production. This will lead in turn to more small firms in the arms business over time;[33] the more vigorous the innovative activity, the greater the reduction in concentration over the medium term.[34] On the whole, process innovation tends to increase concentration, while product innovation tends to decrease it.[35] Innovation in the arms industry today shows both types of innovation: new products are constantly being developed to support the requirements of network-centric warfighting, and process improvements are increasingly being sought in the construction of platforms. Thus, the ultimate industrial effects of this network-centric revolution are still uncertain, but they are likely to produce structural turbulence. Divergent economic trends in closely related sectors of the global arms industry will mean room enough for multiple types of firms, and for multiple types of firms to be successful. Small firms will be needed for component-wise innovation, to produce leading-edge materials, and to develop novel solutions to vexing problems. Larger firms are needed to integrate multiple systems onto the more complex platforms and systems between platforms. Firms with wider scope—though not necessarily the largest scale—will be needed to manage the integration of systems of systems, as metasystems management and systems integration are two distinct disciplines.[36] Bigger will continue to be useful, but it will not always be better.

Notes

Chapter One

1. "To Dissolve, to Disappear," *The Software Revolution: A Survey of Defence Technology*, supplement to the *Economist* (10 June 1995).

2. See John Kenneth Galbraith, *American Capitalism: The New Countervailing Power* (New York: Houghton Mifflin, 1952).

3. Martin Campbell-Kelly, *From Airline Reservations to Sonic the Hedgehog: A History of the Software Industry* (Cambridge: MIT Press, 2004).

4. Roy Rothwell, "The Role of Small Firms in the Emergence of New Technologies," in *Design, Innovation, and Long Cycles in Economic Development*, ed. Christopher Freedman (London: Frances Pinter, 1986), 231–248, see 234, 239.

5. Brian J. L. Berry, *Long-Wave Rhythms in Economic Development and Political Behavior* (Baltimore: Johns Hopkins University Press, 1991), 183.

6. H. G. Barkema and F. Vermeulen, "International Expansion through Start-up or Acquisition: A Learning Perspective," *Academy of Management Journal* 41, no. 1 (Feb. 1998): 7–26; J. L. Bower, "Not All M&As Are Alike—And That Matters," *Harvard Business Review* 79, no. 3 (April 2001): 93–101.

7. W. M. Cohen and D. A. Levinthal, "Innovation and Learning: The Two Faces of R&D," *Economic Journal* 99, no. 397 (Sept. 1989): 569–596; W. M. Cohen and D. A. Levinthal, "Absorptive Capacity: A New Perspective on Learning and Innovation," *Administrative Science Quarterly* 35, no. 1 (1990): 128–152.

8. David R. King and John D. Driessnack, "Investigation the Integration of Acquired Firms in High–Technology Industries: Implications for Industrial Policy," *Acquisition Review Quarterly* (Summer 2003): 263.

9. M. J. Berry and J. H. Taggart, "Combining Technology and Corporate Strategy in Small High-Tech Firms," *Research Policy* 26, nos. 7–8 (April 1998): 833–846.

10. A. R. Fusfield, "Formulating Technology Strategies to Meet the Global Challenges of the 1990s," *International Journal of Technology Management* 4, no. 6 (1989): 601–612.

11. M. Dodgson and R. Rothwell, "Technology Strategies in Small and Medium-sized Firms," in *Technology Strategy and the Firm: Management and Public Policy*, ed. M. Dodgson (Longman, 1989); and M. M. J. Berry, "Technical Entrepreneurship, Strategic Awareness and Corporate Transformation in Small High-tech Firms," *Technovation* 16, no. 9 (Sept. 1996): 487–498.

12. See Stephen Peter Rosen, *Winning the Next War: Innovation and the Modern Military* (Ithaca: Cornell University Press, 1991), 39–43. Others find these political-economic cycles to be at the root of larger, longer, world systems cycles of hegemonic domination. This is an entertaining thought but not one very useful in the medium term. Various incarnations of this idea are found in George Modelski and William R. Thompson, *On Global War: Historical-Structural Approaches to World Politics* (Columbia: University of South Carolina Press, 1988); Joshua Goldstein, *Long Cycles: Prosperity and War in the Modern Age* (New Haven: Yale University Press, 1991); and George and Meredith Friedman, *The Future of War: Power, Technology and American World Dominance in the Twenty-first Century* (New York: Crown Publishers, 1996).

13. E. Mansfield, *The Economics of Technological Change* (New York: Norton, 1968), 215–217.

14. J. W. Markham, "Market Structure, Business Conduct and Innovation," *American Economic Review* (May 1965): 327.

15. See Lee Gomes, "Industry Wise Men Don't Always Produce the Hottest Software," *Wall Street Journal*, 27 Oct. 2003, B1; and Stephen Segaller, *Nerds 2.0.1: A Brief History of the Internet* (New York: TV Books, 1998).

16. Raphael Kaplinski, "Technological Revolution and the Restructuring of Trade Promotion: Some Implications for the Western Middle Powers and the Newly Industrialized Countries," in *Middle Power Internationalism: The North-South Dimension*, ed. Cranford Pratt (Montreal: McGill-Queen's University Press, 1990), 29, n. 21.

17. Gail Kaufman, "Putting the 'A' in F/A-22," *Defense News*, 8 March 2004, 8.

18. The concepts here are drawn from a variety of sources, including Zoltan J. Acs and D. B. Audretsch, "Innovation, Market Structure, and Firm Size," *Review of Economics and Statistics* 69, no. 4 (Nov. 1987): 567–74; and Marianna Mazzucato, *Firm Size, Innovation, and Market Structure: The Evolution of Industry Concentration and Instability*, New Horizons in the Economics of Innovation (Cheltenham: Edward Elgar, 2000), 17. Mazzucato notes that these observations were developed theoretically by William J. Abernathy and Kenneth Wayne in "Limits to the Learning Curve," *Harvard Business Review* 52, no. 5 (Sept.–Oct. 1974): 109–120; and by B. H. Klein in *Dynamic Economics* (Cambridge: Harvard University Press, 1977). The theoretical results have been confirmed by empirical findings in William J. Abernathy and James M. Utterback, "A Dynamic Model of Product and Process Innovation," *Omega* 3, no. 6 (Dec. 1975): 424–441; Acs and Audretsch, "Innovation, Market Structure, and Firm Size"; and Steven Klepper, "Exit, Entry, Growth, and Innovation over the Product Life Cycle," *American Economic Review* 86, no. 3 (June 1996): 562–583.

19. Whether or not it is essential is another question.

20. Comments by Rhett Flater, director of the American Helicopter Society International, in William Matthews, "U.S. Army Urged to Speed Helicopter Plans," *Defense News*, 8 March 2004.

21. Daniel Michaels, "Helicopters Soar Overseas: U.S. Industry Loses Clout to Aggressive European Competitors," *Wall Street Journal*, 2 Jan. 2004, A7.

22. Kurt Hoffman and Raphael Kaplinsky, *Driving Force: The Global Restructuring of Technology, Labor, and Investment in the Automobile and Components Industries* (Greenview: Westview Press, 1988).

23. Michael Spence, "The Learning Curve and Competition," *Bell Journal of Economics* 12 (1981): 49–70.

24. Megan Scully, "The Comanche Dividend: How the U.S. Army Will Spend Its Savings," *Defense News*, 1 March 2004. MD Helicopters was the subject of much speculation in late 2004

and early 2005 as Boeing and Sikorsky were rumored to be teaming up to invest in the company. The two larger helicopter manufacturers wanted to join forces in their pursuit of the Army's Armed Reconnaissance Helicopter (ARH) contract, but also wanted to utilize the technologies that they had jointly developed in the Comanche development. At the same time, they realized that the Army's relative enthusiasm for the Little Bird would make MD an excellent vehicle for their ambitions. MD was subsequently purchased by the private equity fund Patriarch Partners; its investment has not panned out as well as was initially hoped. MD was not successful in securing either the ARH nor the contract in the subsequent Light Utility Helicopter (LUH) program.

25. Peter J. Klenow, "Learning Curves and the Cyclical Behavior of Manufacturing Industries," *Review of Economic Dynamics* 1 (1998): 531–550. My thanks to Leonard Shapiro for some clarifying insights into this question.

26. Giovanni Dosi, "Technological Paradigms and Technological Trajectories: A Suggested Interpretation of the Determinants of Technical Change," *Research Policy* 2, no. 3 (1982): 147–62.

27. Andrew Latham, "Military-Technical Revolution: Implications for the Defence Industry," *Canadian Defence Quarterly* 24, no. 4 (Summer 1995): 19–20.

28. Observation by an anonymous aide to French president Jacques Chirac in J. Fitchett, "Now France, Too, Spies Aloft in a Challenge to American Supremacy," *International Herald Tribune*, 5 July 1996, 12.

29. Latham, "Military-Technical Revolution," 20.

30. Comment by Ken Harris, Managing Director of Australian Defence Industries (ADI), in Damien Kemp, "Privatization, Local Plans Dominate Arena," *Jane's Defence Weekly* (7 April 1999).

31. *Defense Industry: Trends in DOD Spending, Industrial Productivity, and Competition*, GAO/PEMD-97–3 (Washington, D.C.: U.S. General Accounting Office, 1997), 4.

32. Latham, "Military-Technical Revolution," 20.

33. Earll M. Murman, Myles Walton, and Eric Rebentisch, *Challenges in the Better, Faster, Cheaper Era of Aeronautical Design, Engineering, and Manufacturing*, Lean Aerospace Initiative Report RP00–02, Massachusetts Institute of Technology, September 2000, 6.

34. Oliver E. Williamson, *The Economic Institutions of Capitalism: Firms, Markets, Relational Contracting* (New York: Free Press, 1985), 83.

35. Edward Krubasik and Hartmut Lautenschlager, "Forming Successful Strategic Alliances in High-Tech Businesses," in *Collaborating to Compete: Using Strategic Alliances and Acquisitions in the Global Marketplace*, ed. Joel Bleeke and David Ernst (New York: John Wiley & Sons, 1993).

36. Yves L. Doz and Gary Hamel, *Alliance Advantage: The Art of Creating Value through Partnering* (Boston: Harvard Business School Press, 1998), 63

37. J. Lynn Lunsford, "For Pratt, a Cutting-Edge Plane Offers a Shot at a Comeback," *Wall Street Journal*, 12 Dec. 2003, A1.

38. B. Klein, R. G. Crawford, and A. A. Alchian, "Vertical Integration, Appropriable Rents, and Competitive Contracting Process," *Journal of Law and Economics* 21 (Oct. 1978): 300.

39. Bart Nooteboom, *Inter-Firm Alliances: Analysis and Design* (London: Routledge, 1999), 84.

40. Doz and Hamel, *Alliance Advantage*, 208.

41. Alexander Gerybadze, *Strategic Alliances and Process Redesign: Effective Management and Restructuring of Cooperative Projects and Networks* (Berlin: Walter de Gruyter, 1995), 39.

42. Ibid., 29–33, explains the validity of this framework.

43. For a fascinating study of how the U.S. Army encountered this problem recurringly in designing a troop carrier, see W. Blair Haworth, *The Bradley and How It Got That Way: Technology, Institutions, and the Problem of Mechanized Infantry in the United States Army*, Contributions in Military Studies, 180 (Westport, CT: Greenwood Press, 1999).

44. W. G. Ouchi and M. K. Bolton, "The Logic of Joint Research and Development," *California Management Review* 30, no. 1 (Spring 1988): 9–33.

45. The concept was first articulated by the noted British economist Alfred Marshall (1842–1924). See Graham Bannock, Ron Baxter, and Evan Davis, *The Economist Dictionary of Economics* (Princeton: Bloomberg Press, 2003), 320–321.

46. A. A. Alchian, "Specificity, Specialization, and Coalitions," *Journal of Institutional and Theoretical Economics* 140, no. 1 (1984): 36.

47. Nooteboom, *Inter-Firm Alliances*, 82.

48. A. Madhok and S. B. Tallman, "Resources, Transactions, and Mergers: Managing Value through Inter-Firm Collaborative Relationships," *Organization Science* 9, no. 3 (1998): 326–339.

49. J. Hagedorn, "Understanding the Rationale of Strategic Technology Partnering: Interorganizational Modes of Cooperation and Sectoral Differences," *Strategic Management Journal* 14 (1993): 371–386; and D. C. Mowery, J. E. Oxley, and B. S. Silverman, "Strategic Alliances and Interfirm Knowledge Transfer," *Strategic Management Journal* 17 (1996): 77–91.

50. Gerybadze, *Strategic Alliances and Process Redesign*, 130.

51. Most of the sectoral standards are defined by revenue, but in most aerospace and arms sectors, the standard is one of staff. The question is not trivial: if revenue is the measure, then the degree to which a firm vertically integrates production will not affect its status, which is useful from an industrial structure standpoint. On the other hand, if the standard is one of staff size, then the degree of vertical integration will affect status; this will be useful in organizational studies. Regardless, the SBA's standards are available at www.sba.gov/size.

52. Stephen Goldsmith, "Can Business Really Do Business with Government?" *Harvard Business Review* (May–June 1997): 110–121.

53. Steven Van Evera, *Qualitative Methods in Political Science* (Ithaca: Cornell University Press, 1997). For more recommendations for the focused, comparative case study approach, see Alexander L. George, "Case Studies and Theory Development: The Method of Structured, Focused, Comparison," in *Diplomacy: New Approaches in History, Theory and Policy*, ed. Paul G. Lauren (New York: Free Press, 1979), 43–68; Charles Edquist, introduction to *Systems of Innovation: Technologies, Institutions, and Organizations*, ed. Charles Edquist (London: Pinter, 1997), 1–35; Vasudevan Ramanujam and P. Varadarajan, "Research on Corporate Diversification: A Synthesis," *Strategic Management Journal* 10 (1989): 523–551; and Robert K. Yin, *Case Study Research: Design and Methods*, 2nd ed. (Thousand Oaks, CA: Sage, 1994).

Chapter Two

1. David Shook, "Spectrum Astro: The Rising Star of Aerospace," *Business Week*, 4 May 2001.

2. Northrop Grumman chairman Kent Kresa, briefing to the Fifteenth Annual Solomon Smith Barney Global Industrial Manufacturing Conference, New York City, 27 Feb. 2002.

3. As mentioned above, whatever one thinks of the theories of Admiral Bill Owens, the former Vice Chairman of the Joint Chiefs of Staff, his "systems of systems" term is apt. For his views on the future of war, see William Owens and Ed Offley, *Lifting the Fog of War* (New York: Farrar Straus & Giroux, 2000).

4. Vago Muradian, "Few Northrop Grumman-TRW Antitrust Issues Seen; Some in DoD Focus on Debt, Programs," *Defense Daily International* 3, no. 17 (1 March 2002).

5. "Spectrum Astro/Northrop Grumman Complete SBIRS Low Review," *Space Daily*, 7 May 2001. System design review (SDR) is one of several formal checkpoints for a contractor team engaged in a program definition and risk reduction phase of work for the Department of Defense. It follows the preliminary design review (PDR) and precedes the critical design review (CDR). At each stage, the design of the system is expected to be more mature and subject to less schedule and cost risk. In the case of SBIRS Low, that might not be obvious to an outside observer.

6. This view has been prevalent in the literature for some time. See, for example, P. Mariti and R. H. Smiley, "Cooperative Agreements and the Organization of Industry," *Journal of Industrial Economics* 31 (1983): 437–451; and P. Ghemawat, M. E. Porter, and R. A. Rawlinson, "Patterns of International Coalition Activity," in *Competition in Global Industries*, ed. M. E. Porter (Boston: Harvard Business School Press, 1986). Mergers and acquisitions, on the other hand, are probably at least as often justified on the basis of cost savings or pricing power.

7. The technical details of the system are indeed important to understanding the managerial and economic challenges that it presented. Mastery of these details has eluded at least a few would-be analysts. For example, in "The Trials and Tribulations of SBIRS-Low," David Grahame of the British-American Security Information Council (BASIC) writes that "consisting of 24–30 satellite radars [?] in Low Earth Orbit, the SBIRS-Low system is intended to track hostile ICBMs as they travel through space." If one does not recognize the satellites as infrared sensors, not much more can be intelligently said about them.

8. Analysis provided by Prof. Michael R. Rip, James Madison College, Michigan State University.

9. For a discussion of the limitations of LANTIRN pods in this role, see M. R. Rip and J. M. Hasik, *The Precision Revolution: GPS and the Future of Aerial Warfare* (Annapolis: Naval Institute Press, 2002).

10. Additional analysis provided by Michael R. Rip. There is some dispute about the exact number of launches and launchers—these figures include those missiles that flew into the sea but exclude the untested, homemade Iraqi launch vehicles that proved incapable of producing successful launches.

11. "Space-based Infrared Systems: The First Step in a Credible Missile Defense," undated and unattributed briefing by the SBIRS Program Office, Los Angeles AFB.

12. Amy McAuliffe, "Infrared Focal Plane Arrays Drive Design of New Space Surveillance System,' *Military and Aerospace Electronics* 6, no. 5 (May 1995): 1.

13. Under Secretary of Defense for Acquisition, Technology and Logistics Pete Aldridge, special Pentagon briefing to update Department of Defense acquisition programs, 21 Dec. 2001.

14. "Air Force Under Secretary Could Seek Alternative for SBIRS High," *Satellite Week*, 18 Feb. 2002. Teets's criticism of prime contractor Lockheed Martin is particularly stinging as he is a former Lockheed Martin chief operating officer.

15. See John R. Ward, *Beyond "Integrated Weapon System Management": Acquisition in Transition* (Washington, D.C.: Industrial College of the Armed Forces, National Defense University, 1993).

16. Vago Muradian, "Aldridge: DoD to Play Greater Industrial Role in Profits, Programs, Prime Power," *Defense Daily International*, 22 March 2002.

17. Richard J. Fickes and Kenneth A. Good, "Space-Based Infrared System—Supportability Engineering and Acquisition Reform in an Existing Acquisition Environment," *Air Force Journal of Logistics* 23, no. 1 (Spring 1999): 22–28.

18. Anne Marie Squeo, "Crucial Lockheed Satellite System for Use in Missile Defense Is over Cost and Late," *Wall Street Journal*, 16 Nov. 2001, 2.

19. For more on the innovations of the NPOESS program, see Robert Graham, "The Transformation of Contract Incentive Structures," *Acquisition Review Quarterly* 10, no. 3 (Summer 2003): 237.

20. Stephen Barlas, "Troubled Weather Satellite Program," *IEEE Spectrum* (June 2006).

21. James H. Soloman et al., *Space-Based Infrared System-Low at Risk of Missing Initial Deployment Date* (Washington, D.C.: U.S. General Accounting Office, Feb. 2001).

22. Congressional testimony by Under Secretary of Defense for Acquisition and Technology Jacques Gansler, 28 June 2000.

23. "Pentagon to Delay Deployment of SBIRS Satellites," Armed Forces Newswire Service, 12 Jan. 1999.

24. *Defense Daily*, 26 Feb. 1999.

25. "Air Force Cancels SBIRS-Low," Armed Forces Newswire Service, 9 Feb. 1999. The headline was a bit overreaching.

26. According to General Ralph Eberhardt, USAF, Commander-in-Chief, Space Command. Kerry Gildea, "Eberhardt Reports Top Goal of Getting SBIRS Back on Track," *Defense Daily* 213, no. 50 (15 March 2002).

27. Robert Wall, "New Space-Based Radar Shaped by SBIRS Snags," *Aviation Week and Space Technology* 156, no. 7 (18 Feb. 2002): 30.

28. Information from the TRW-Raytheon team Web site at http://www.sbirslowteam.com (accessed in March 2002).

29. "Lockheed Martin and the Boeing Company Join [*sic*] Spectrum Astro/Northrop Grumman SBIRS Low Team," press release 1139-PR-U22963, Spectrum Astro Inc., Gilbert, Arizona, 20 March 2001.

30. Oliver E. Williamson, *The Economic Institutions of Capitalism: Firms, Markets, Relational Contracting* (New York: Free Press, 1985), 83.

31. Kerry Gildea, "SBIRS Low Team Competitors Move into Home Stretch," *Defense Daily*, 9 April 2001.

32. A. A. Alchian, "Specificity, Specialization, and Coalitions," *Journal of Institutional and Theoretical Economics* 140, no. 1 (1984): 36.

33. "Senior Defense Official," Pentagon briefing on current acquisition programs, 4 Feb. 2002.

34. "Fact Sheet: Space Tracking and Surveillance System (STSS)," U.S. Missile Defense Agency, Oct. 2002.

35. Wade Boese, "Pentagon Asks for $9.1 Billion in Missile Defense Funding," *Arms Control Today* (March 2003).

36. Jeremy Singer, "Pentagon Renames SBIRS Low," *Space News*, 18 Dec. 2002.

37. "TRW/Aerojet-Built Defense Support Program Satellite Slated for Launch on July 27 from a Titan IVB," Business Wire, 25 July 2001.

38. Ibid.

39. Pierre Dussauge, "Alliances et cooperations dans l'aerospatial et armement: Bilan et perspectives dans le context de l'après-guerre froide," *Economie appliquée* 46, no. 3 (1993): 117–153.

40. K. Hartley and S. Martin, "International Collaboration in Aerospace," *Science and Public Policy* 17, no. 3 (1990): 143–151.

41. See Roger E. Bilstein, *The American Aerospace Industry: From Workshop to Global Enterprise* (New York: Twayne Publishers, 1995), 120.

42. Kerry Gildea, "TRW Restructuring to Secure Position in Missile Defense Arena," *Defense Daily International* 2, no. 19 (15 March 2002).

43. "Aerojet Joins TRW-Raytheon Team in U.S. Air Force Missile-Defense Program," *Satellite Today*, 23 Nov. 1999; see also 17 Aug. 1999.

44. Shook, "Spectrum Astro." Rival Orbital Sciences was founded by (no relation) David W. Thompson. For a comparison, see Jill Hecht Maxwell, "Rising, Falling, Rising Star," *Inc*, 15 Nov. 2001.

45. The terms of the acquisition were not announced as Spectrum Astro was privately held.

46. Taylor Dinerman, "General Dynamics Buys Spectrum Astro: Good Deal for Whom?" *Space Review*, 29 March 2004.

47. See Jonathan Karp and Andy Pasztor, "Can Defense Contractors Police Their Rivals Without Conflicts?" *Wall Street Journal*, 28 Dec. 2004.

Chapter Three

1. Senior Bush administration official in December 2001 quoted in Joseph Fitchett, "High-Tech Weapons Change the Dynamics and the Scope of Battle War in the Computer Age," *International Herald Tribune*, 28 Dec. 2001.

2. This story is drawn from a number of sources, including David Ensor, "U.S. Kills *Cole* Suspect; CIA Drone Launched Missile," CNN.com, 5 Nov. 2002; David Ensor, "American Killed in U.S. Strike in Yemen; U.S. Official: 'It doesn't change anything,'" Cable News Network, 8 Nov. 2002; Philip Smucker and Toby Harnden, "CIA Missile Team Stalked bin Laden's Top Man for Months," *Daily Telegraph*, 6 Nov. 2002; Faye Bowers, "U.S. Pulls Out New Tools, New Rules: CIA Slaying by Drone Signals Bush's Resolve to Hunt Down al Qaeda, Even By Controversial Means," *Christian Science Monitor*, 6 Nov. 2002; Elaine Monaghan and Daniel McGrory, 'Death of Terror Chief Deals Severe Blow to al-Qaeda," *Times*, 5 Nov. 2002; James Risen and Judith Miller, "U.S. Is Reported to Kill Al Qaeda Leader in Yemen," *The New York Times*, 5 Nov. 2002; Walter Pincus, "Missile Strike Carried Out with Yemeni Cooperation: Official Says Operation Authorized Under Bush Finding," *Washington Post*, 6 Nov. 2002, A10; Daniel McGrory, Michael Evans, and Elaine Monaghan, "Robotic Warfare Leaves Terrorists No Hiding Place: A New Era of Combat Has Dawned, Where Death Is Delivered by an Unseen Hand from Hundreds of Miles Away," *Times*, 6 Nov. 2002; and David Johnston and David E. Sanger, "Yemen Killing Based on Rules Set Out by Bush," *New York Times*, 6 Nov. 2002. Note that one of the other men killed was Ahmed Hijazi, a U.S. citizen. While the CIA had not known in advance that he was in the car, his death was not exactly a source of regret.

3. Information courtesy of Lt. Colonel Stan Coerr, USMCR, and Darren Bradley, formerly of General Atomics.

4. Sue Baker, "Predator Missile Launch Test Totally Successful," *Air Force Print News*, 28 Feb. 2001.

5. Michael Evans, "Spy Plane Now CIA's Deadliest Weapon," *Times*, 5 Nov. 2002; Walter Pincus, "U.S. Strike Kills Six in Al Qaeda: Missile Fired by Predator Drone; Key Figure in Yemen Among Dead," *Washington Post*, 5 Nov. 2002, A1; Walter Pincus, "Tenet Says Al Qaeda Still Poses Threat: CIA Drone Fires Missiles at Possible Terror Leaders," *Washington Post*, 7 Feb. 2002, A1.

6. Richard M. Clark, *Uninhabited Combat Aerial Vehicles: Airpower by the People, for the People, But Not with the People* (USAF School of Advanced Airpower Studies, 1999), 18–19.

7. See, for example, Michael R. Rip and Joseph Fonatanella, "A Window on the Arab-Israeli 'Yom Kippur' War of October 1973: Military Reconnaissance from High Altitude and Space," *Intelligence and National Security* 6, no. 1 (1991): 57; and Michael R. Rip, "Military Photo-Reconnaissance during the Yom Kippur War: A Research Note," *Intelligence and National Security* 7, no. 2 (1992): 126.

8. Gerald H. Turley, "Electronic Battlefields Are Here," *Proceedings of the U.S. Naval Institute* (Nov. 1982): 110.

9. W. Seth Carus, "The Bekaa Valley Campaign," *Washington Quarterly* (Autumn 1982): 39–41.

10. Two programs were launched during the Reagan administration, but these were cancelled during the George H. W. Bush administration by then USAF Chief of Staff General Merrill McPeak. This may have been because the USAF leadership had decided to fob the reconnaissance mission off on the National Reconnaissance Office, in some ways a subsidiary of the USAF anyway. Thomas P. Ehrhard, "Unmanned Aerial Vehicles: A Comparative Study of Weapon System Innovation" (Ph.D. diss., Johns Hopkins University, 2000), 513.

11. Benjamin F. Schemmer, "Where Have All the RPVs Gone?" *Armed Forces Journal International* (Feb. 1982): 38.

12. Comments to the U.S. Congress by Under Secretary of the Air Force James W. Plummer, reported in "Temper RPV Enthusiasm," *Aviation Week and Space Technology* (23 June 1975): 7. UAVs at the time, of course, were called "remotely piloted vehicles" (RPVs)—the change in terminology reflects the lessened requirement for operator oversight in new aircraft such as the RQ-4 Global Hawk.

13. Vigorous requirements management for affordability was one of the primary recommendations of the Defense Science Board's UAV Task Force in 1997. See Kenneth Israel and Robert Nesbit, *Defense Science Board Study on Unmanned Aerial Vehicles and Uninhabited Combat Air Vehicles* (Office of the Under Secretary of Defense for Acquisition, Technology, and Logistics, 2004).

14. "U.S. Buys Israeli Pilotless Planes," *New York Times*, 24 May 1984; and Steven L. Spiegel, "U.S. Relations with Israel: The Military Benefits," *Orbis* (Fall 1986): 486.

15. David A. Fulghum, "CIA to Fly Missions from inside Croatia," *Aviation Week and Space Technology* (11 July 1994): 20.

16. Michael Barzelay and Colin Campbell, *Preparing for the Future: Strategic Planning in the U.S. Air Force* (Washington, D.C.: Brookings Institution Press, 2003), 157.

17. Fulghum, "CIA to Fly Missions from Inside Croatia," 20; and Tim Ripley, "UAVs in Action: 1964 to 2002 (Part 2)," *Air International* (June 2002): 350.

18. "Gnats Weathered Out," *Aviation Week and Space Technology* (14 Feb. 1994): 19.

19. "Spying on Bosnia," *Aviation Week and Space Technology* (6 June 1994): 23; and David A. Fulghum and John D. Morrocco, "U.S. Military to Boost Tactical Recon in 95," *Aviation Week and Space Technology* (9 Jan. 1995): 22.

20. "Stealthy UAV Is Flying Wing," *Aviation Week and Space Technology* (11 July 1994): 21.

21. Michael R. Thirtle, Robert V. Johnson, and John L. Birker, *The Predator ACTD: A Case Study for Transition Planning to the Formal Acquisition Process*, MR-899 (Santa Monica: RAND, 1997).

22. Memorandum from John Deutch, Under Secretary of Defense for Acquisition, to the Assistant Secretary of the Navy for Research, Development, and Acquisition, "Endurance

Unmanned Aerial Vehicle (UAV) Program," 12 July 1993. A national imagery interpretability rating of 6 requires a system to be able to, among other things, "identify the spare tire on a medium-sized truck." See L. A. Maver, C. D. Erdman, and K. Riehl, "Imagery Interpretability Rating Scales," *Society for Information Display* (1995).

23. Stacey Evers, "Gnat-750 May Raise Profile of Military UAVs," *Aviation Week and Space Technology* (7 Feb. 1994): 54.

24. Alexander Nicoll, "Robot Weapons Convince Skeptics," *Financial Times*, 5 Feb. 2000: 3.

25. David A. Fulghum and John D. Morrocco, "CIA to Deploy UAVs to Albania," *Aviation Week and Space Technology* (31 Jan. 1994): 20–22.

26. Michael R. Thirtle, Robert V. Johnson, and John L. Birker, *The Predator ACTD: A Case Study for Transition Planning to the Formal Acquisition Process*, MR-899 (Santa Monica: RAND, 1997).

27. Fulghum and Morrocco, "CIA to Deploy UAVs to Albania."

28. David A. Fulghum, "Predators Bound for Bosnia Soon," *Aviation Week and Space Technology* (13 Nov. 1995): 73.

29. Sue Baker, "Predator Missile Launch Test Totally Successful," *Air Force Print News*, 28 Feb. 2001.

30. Adam J. Hebert, "Compressing the Kill Chain: The Goal Is to Put Weapons on Time-Sensitive Target in 'Single-Digit' Minutes," *Air Force Magazine* 86, no. 3 (March 2003).

31. Information from Darren Bradley, former export licensing manager for General Atomics.

32. Michael Evans, "Predator Is the Soldier's Dream Machine," *Times*, 6 Nov. 2002.

33. Brendan P. Rivers, "Predator Launches Mini-UAV," *Journal of Electronic Defense* (Oct. 2002): 37.

34. "Predator to Make Debut over War-torn Bosnia," *Aviation Week and Space Technology* (10 July 1995): 47.

35. "Imagery from Bosnia Expected to Improve," *Aviation Week and Space Technology* (31 July 1995): 20.

36. Tim Ripley, "U.S. UAVs over the Balkans," *Air Forces Monthly* (Oct. 2001): 60; "Two Predators Destroyed in Bosnia," *Aviation Week and Space Technology* (21 Aug. 1995): 24.

37. Ripley, "UAVs in Action: 1964 to 2002 (Part 2)," 351.

38. "Send in the Drones," *Economist* (10 Nov. 2001): 73.

39. Anthony H. Cordesman, *The Lessons and Non-Lessons of the Air and Missile Campaign in Kosovo* (Westport, CT: Praeger, 2001), 352–354.

40. Tim Ripley, "UAVs over Kosovo: Did the Earth Move?" *Defense Systems Daily*, 1 Dec. 1999; and Rip and Hasik, *Precision Revolution*, 407–411.

41. Tim Ripley, "British Army to Rationalise UAV Operations," *Jane's Defence Weekly* (12 Sept. 2001): 18.

42. *Defense Weekly*, daily update, 3 July 2002.

43. "Unmanned U.S. Plane Reported Down in Pakistan," Associated Press, 22 Jan. 2002; and Jim Skeen, "Global Hawk Crashes," *Los Angeles Daily News*, 1 Jan. 2002.

44. David A. Fulghum, "More UAVs Shift to Afghan Duty," *Aviation Week and Space Technology* 155, no. 18 (29 Oct. 2001): 44.

45. David A. Fulghum, "Uptown, Downtown, Predators Are Striking in the Heart of Baghdad and Patrolling the Western Wastelands of Iraq," *Aviation Week and Space Technology* 158, no. 14 (7 April 2003): 25. The USAF began thinking about adding an air-to-air capability to the Predator not later than its successful tests with the Hellfire in early 2001. See Marc Strass, "Air

Force Looking to Demonstrate Predator Air-to-air Capability," *Defense News*, 22 March 2001. Taking on a MiG-25, however, was quite a stretch for a piston-driven aircraft.

46. The USAF also lost two Predators to Iraqi ground fire in August and September 2001. See Craig Hoyle, "U.S. Build-up Highlights UAV Shortage," *Jane's Defence Weekly* (10 Oct. 2001): 5.

47. David A. Fulghum, "Unmanned Reconnaissance Aircraft Was Used in Iraq: Stealth UAV Goes to War," *Aviation Week and Space Technology* (6 July 2003).

48. Testimony by Dyke Weatherington, deputy head of the Pentagon's UAV planning task force, before the Armed Services Committee of the U.S. Congress, cited in Marc Selinger, "U.S. Using More Than 10 Types of UAVs in Iraq War, Official Says," *Aerospace Daily*, 27 March 2003.

49. Nicholas Rufford, "Fly-away Drones Put Robot Air Force Plans Off Course," *Sunday Times*, 22 June 2003.

50. Francis Harris, "In Las Vegas a Pilot Pulls the Trigger. In Iraq a Predator Fires Its Missile," *Telegraph*, 2 June 2006.

51. Tom Kington, "Italian Predator Crashes in Iraq," *Defense News*, 17 May 2006, and idem, "Crashed Italian Predator Flies Again," *Defense News*, 5 July 2006.

52. Gail Kaufman, "USAF Predator's Mission Continues in Iraq," *Defense News*, 8 March 2004; and Bob Thompson, "Predator Unit Boasts 20 Unmanned Aircraft, Iraq Covered," Air Combat Command News Service, 6 July 2006.

53. Bing West, *No True Glory: A Frontline Account of the Battle for Fallujah* (New York: Bantam, 2005).

54. Michael Barzely and Colin Campbell, *Preparing for the Future: Strategic Planning in the U.S. Air Force* (Washington: Brookings Institution Press, 2003), 156.

55. David A. Fulghum, "Anti-Air Defense Role Eyed for Predator," *Aviation Week and Space Technology* (22 June 1998): 61–62.

56. "Predator Fleet to Expand," USAF press release, 18 March 2005; and "General Atomics Aeronautical Systems Expands Production Capability," General Atomics press release, 14 March 2005. At the time, GA was expanding its floor space from 640,000 to 800,000 square feet. While this seems a large factory floor for a relatively small aircraft, the company actually produced roughly half its own parts on-site. This remains a relative rarity in the aircraft industry.

57. Glen Goodman, "U.S. Air Force to Expand Predator Fleet," *C4ISR Journal*, 13 April 2005.

58. Anthony M. Cordesman, *The Lessons of Afghanistan: War Fighting, Intelligence, and Force Transformation* (Washington, D.C.: CSIS Press, 2002), 104–105.

59. W. Michael Riesman, "The Lessons of Qana," *Yale Journal of International Law* 22 (1997): 396–397. For more on the legal wrangling, see Anthony J. Lazarski, "Legal Implications of the Uninhabited Combat Air Vehicle," *Aerospace Power Journal* (Summer 2002).

60. *New York Times*, 7 July 2002, 8; *Washington Post*, 9 July 2002, 17.

61. It is interesting to note that U.S. federal law prohibits this activity for long-distance truck drivers but permits it for fighter-bomber pilots. See "Pills Cited in Mistaken Afghan Bombing," Associated Press, 14 Jan. 2003; and "Friendly Fire Pilots: Air Force Pushes Go Pills," Cable News Network, 2 Jan. 2003.

62. Michael Sirak, "Interview: James Roche, Secretary of the U.S. Air Force," *Jane's Defence Weekly* (9 Jan. 2002): 32.

63. David Persampieri, "Military Feels Bandwidth Squeeze as the Satellite Industry Sputters," *Wall Street Journal*, 10 April 2002, A1.

64. Comment by Lieutenant General Ronald Keys, USAF deputy chief of staff for air and space operations, in Joris Janssen Lok, "Communications Weaknesses Endanger Allied Integration in U.S.-led Air Campaigns," *Jane's International Defense Review* (March 2004): 4.

65. Adam Hebert, "New Horizons for Combat UAVs," *Air Force Magazine* (Dec. 2003): 72.

66. See the comments by then secretary of the Air Force James Roche in Thom Shanker, "Incentives Added for Pilots of Remote Predator Planes," *New York Times*, 17 Oct. 2002, A18.

67. David A. Fulgham, "Anti-Air Defense Role Eyed for Predator," *Aviation Week and Space Technology* (22 June 1998): 61–62.

68. David Bond, "Paying UAV Pilots," *Aviation Week and Space Technology* 157, no. 17 (21 Oct. 2002): 21.

69. David A. Fulghum, "Air Force Prepares New UAV Acquisitions, Operations," *Aviation Week and Space Technology* (27 Nov. 1995): 54. For more on who should fly drone aircraft, see Julie A. Fenimore, "The Navy Isn't Serious about Using UAVs," *Proceedings of the United States Naval Institute* (Jan. 2002): 94; and Jennifer Palmer, "Pilots Warm Up to Predator Assignments," *Air Force Times*, 10 July 2000, 17–18; and David Bond, "Paying UAV Pilots," *Aviation Week and Space Technology* (21 Oct. 2002): 21.

70. Linda Shiner, "Predator: First Watch—The Air Force's Latest Remotely Operated Reconnaissance Marvel Gets Its First Real-world Test in the Skies over Bosnia," *Air and Space Magazine* (April–May 2001).

71. Earl Odom, "Future Missions for Unmanned Aerial Vehicles: Exploring Outside the Box," *Aerospace Power Journal* (Summer 2002).

72. Such was the impression of retired Major General Kenneth Israel, one of the best-known UAV experts in the service. See Thomas E. Ricks and Anne Marie Squeo, "The Price of Power—Sticking to Its Guns: Why the Pentagon Is Often Slow to Pursue Promising Weapons," *Wall Street Journal*, 12 Oct. 1999, A1.

73. Comment by Major David Kumashiro, a Predator pilot from the 15th Expeditionary Reconnaissance Squadron out of Indian Springs Airfield, Nevada, quoted in Fulghum, "Uptown, Downtown, Predators Are Striking in the Heart of Baghdad," 25.

74. "100th Predator," *Jane's International Defense Review* (Feb. 2004): 25.

75. Ron Laurenzo, "Air Force Has Lost One-third of Its Predators," *Defense Week*, 8 April 2002, 2.

76. Sharon Hobson, "Sperwer Crash Leaves Canadian Peacekeepers without UAV Capability," *Jane's Defence Weekly* (4 Feb. 2004): 12.

77. Jeffrey A. Drezner, Geoffrey Sommer, and Robert S. Leonard, *Innovative Management in the DARPA High Altitude Endurance Unmanned Aerial Vehicle Program: Phase II Experience*, MR-1054-DARPA (Santa Monica: RAND, 1999), xiii.

78. Mackubin Thomas Owens, "Transforming Transformation: Defense-Planning Lessons from Iraq," *National Review* (23 April 2003). See also Williamson Murray and Thomas O'Leary, "Military Transformation and Legacy Forces," *Joint Forces Quarterly* (Spring 2002): 20–27.

79. This experimentation is essential to progress. As former Defense Secretary Donald Rumsfeld put it, "If we waited till we had something perfected, we never would have used the Predator, we never would have used the Predator or the Hellfire in Afghanistan or Iraq and it was a very effective weapon." Comments during a media availability session after a missile defense conference in Huntsville, Alabama, 18 Aug. 2004.

80. Mark Hewish, "Predator B Packs a Substantial Punch," *Jane's International Defense Review* (Feb. 2002): 21.

81. Michael Sirak, "USA Pushes Armed UAV Plans," *Jane's Defence Weekly* (25 Feb. 2004): 9.

82. Nick Cook, "Predator Closes Sensor-to-shooter Gap for USAF," *Jane's Defence Weekly* (13 Feb. 2002): 28–29.

83. Barbara Opall-Rome, "UAVs Playing a Major and Growing Role in Military Operations," *Defense News*, 23 April 2004.

84. See Michael Porter, *The Competitive Advantage of Nations* (New York: Free Press, 1990), 81–86.

85. Timothy D. Hoyt, "Revolution and Counter-Revolution: The Role of the Periphery in Technological and Conceptual Innovation," in *The Diffusion of Military Technology and Ideas*, ed. Emily O. Goldman and Leslie C. Eliason (Palo Alto: Stanford University Press, 2003), 179–201, quotation on 186; Eliot Cohen, Michael Eisenstadt, and Andrew Bacevich, *Knives, Tanks, and Missiles: Israel's Security Revolution* (Washington: Washington Institute for Near East Policy, 1998), 126; *The Future Security Environment*, report to the Pentagon Commission on Integrated Long-Term Strategy, Oct. 1988, 27; and Porter, *Competitive Advantage of Nations*.

86. Isabelle Dostaler, "Product Development Performance: Evidence from the Aerospace Industry," *Academy of Management Proceedings* (2002).

87. Dan Breznitz, *The Military as a Public Space: The Role of the IDF in the Israeli Software Innovation System*, Industrial Performance Center Working Paper 02–005 (Boston: MIT Industrial Performance Center, 2002).

88. See, among other accounts, "Hezbollah Shocks with UAV Capability," *C4ISR Journal* (4 Oct. 2006); and Yaakov Katz, "Hizbullah UAV Shot Down off Acre Coast," *Jerusalem Post*, 7 Aug. 2006.

89. Coverage of this program has been extensive in the trade press. See, for example, Bill Sweetman and Nick Cook, "Hidden Agenda: What Next for Low Observables Technology?" *Jane's Defence Weekly* (20 June 2001): 59; Mark Hewish and Joris Janssen Lok, "Cleared for Take-off?" *Jane's International Defense Review* (Oct. 2001): 65; and Michael Sirak, "UCAV Programme Nears First Flight," *Jane's Defence Weekly* (6 March 2002): 9.

90. Malcolm English, "EADS Technology Forum," *Air International* (Dec. 2001): 366–367.

91. Bradley Graham, "Air Force Analysts Feel Vindicated on Iraqi Drones," *Washington Post*, 26 Sept. 2003, A23.

92. Richard R. Burgess, "Navy Aircraft Managers Refine Requirements in Post-Iraq Era," *Sea Power*, June 2003.

93. Alexander Nicoll, "Afghan Campaign Boosts General Atomics," *Financial Times*, 21 Jan. 2002, 15; and "Robot Weapons Convince Skeptics," *Financial Times*, 5 Feb. 2002, 3.

94. William H. Johnson, "UAV 101," *Proceedings of the U.S. Naval Institute* (Nov. 2001): 91. Actual shipboard use of high-performance unmanned aircraft still has some way to go. For a useful description of the challenges, see Hewish and Lok, "Cleared for Take-off?" 59–69.

95. "Lockheed Martin and General Atomics Demo Integrated Maritime Surveillance," *Space Daily*, 7 May 2004.

96. Richard R. Burgess, "Aircraft Manufacturers Bring Different Capabilities to BAMS Starting Line," *Sea Power* (June 2004).

97. Jason Sherman and Gail Kaufman, "USAF Urges Navy to Join Global Hawk UAV Effort," *Defense News*, 8 March 2004; "$143M for Global Hawk Cost Overruns," *Defense Industry Daily*, 25 April 2005. Adding the development costs, each Global Hawk will have cost closer to $120 million. See Renae Merle, "Price of Global Hawk Surveillance Program Rises," *Washington Post*, 7 Dec. 2004.

98. Gregor Ferguson, "National Air Support to Offer Predator," *Australian Defence Magazine* (Aug. 2004).

99. Zev Stub, "Elbit Group Chosen to Supply UK Spy Drone," *Jerusalem Post*, 20 July 2004.

100. A trademark dispute in the summer of 2006 had led GA to temporarily stop calling the aircraft the Warrior and refer to it merely as the ER/MP.

101. Joshua Kucera, "U.S. Army, Air Force Told to Join Forces on FCA, Warrior," *Jane's Defence Weekly* (20 Jan. 2006).

102. "$8B ACS Spy Plane Program Shot Down by Pentagon," *Defense Industry Daily*, 13 Jan. 2006.

103. See Almarin Philips, *Technology and Market Structure: A Study of the Aircraft Industry* (Lexington, MA: D. C. Heath, 1971).

104. Comment by Larry Dickerson of Forecast International in "Afghan Ops Bolster UAV Market," *Space Daily*, 19 June 2002.

105. Ted McKenna, "Cleared for Action," *Journal of Electronic Defense* (Sept. 2003).

Chapter Four

1. Peter Huber, "The Palm Pilot–JDAM Complex," *Forbes* (May 2003).

2. This story is taken from Phillip O'Connor, "Combat and Compassion: Duty in Afghanistan Tests Elite Green Berets' Skills in Fighting, Building Trust and Reaching Out to War's Victims," *St. Louis Post-Dispatch*, 23 Feb. 2004.

3. O'Connor, "Combat and Compassion," A6.

4. "Space Critical to U.S. Military and Operation Enduring Freedom," *Air Force Print News*, 11 June 2002.

5. Donald MacKenzie, *Inventing Accuracy: A Historical Sociology of Nuclear Missile Guidance* (Boston: MIT Press, 1993).

6. Vernon Loeb, "Bursts of Brilliance: How a String of Discoveries by Unheralded Engineers and Airmen Helped Bring America to the Pinnacle of Modern Military Power," *Washington Post Magazine*, 15 Dec. 2002, 24.

7. George Cahlink, "Birth of a Bomb: The Quest for Smart, Cheap and Versatile Munitions," *Government Executive* 35, no. 11 (Aug. 15, 2003).

8. Information courtesy of the Air Combat Command public affairs office, Langley Air Force Base, Virginia.

9. Ted Carlson, "JDAM and the Cat," *Air Forces Monthly* (April 2002): 60–63; Gert Kromhout, "Tomcat Renaissance," *Air Forces Monthly* (June 2001): 42–47; Denise Deon and Sandra Schroeder, "U.S. Navy's F-14D Tomcats Gain JDAM Capability," press release from Naval Air Systems Command, 20 March 2003; and Hunter Keeter, "Tomcats Mark Milestone over Iraq," *Defense Daily*, 24 March 2003. The F-14 was also upgraded in the late 1990s with a digital flight control system (DFCS) to compensate for its poor handling near stall speeds. The DFCS, incidentally, is a derivative of the system developed for the Eurofighter by GEC Marconi, now part of BAE Systems.

10. "U.S. Navy's F-14D Tomcats Gain JDAM Capability," Naval Air Systems Command press release, 20 March.

11. NATO, *Kosovo/Operation Allied Force After-Action Report: Report to Congress* (Washington, D.C.: Department of Defense, 2000), 96; and Anthony H. Cordesman, *The Lessons and Non-Lessons of the Air and Missile Campaign in Kosovo* (Westport, CT: Praeger, 2001), 270.

12. This figure comes from a source in production at Boeing, who commented that the number was far in excess of what they had expected. Noted military analyst William Arkin compiled a quite precise figure of 4,613 JDAMs used. See Anthony H. Cordesman, *The Lessons of Afghanistan: War Fighting, Intelligence, and Force Transformation* (Washington: CSIS Press, 2002), 7.

13. "Boeing Co. JDAM Most Widely Used Precision Bomb in Afghanistan," *Bloomberg News*, 10 Dec. 2001.

14. Carlos Tapia, "Ammo Provides OEF Firepower," Air Combat Command News Service, 12 March 2002.

15. David A. Fulghum and Robert Wall, "Heavy Bomber Attacks Dominate Afghan War: New Real-time Targeting, Plus Long Endurance, Has Recast the Bomber Fleet as a Full-time Battlefield Menace," *Aviation Week and Space Technology* (3 Dec. 2001).

16. Benjamin S. Lambeth, *Air Power against Terror: America's Conduct of Operation Enduring Freedom* (Santa Monica: RAND, 2005), 79.

17. Vernon Loeb, "From U.S., Bat-Winged B-2 Strikes at Taliban; Bomber Aloft for Days in Long-Distance Run," *Washington Post*, 20 Oct. 2001, A17.

18. Tim Dougherty, "B-1 Is Tailor-made for Operation Enduring Freedom,' Air Force Print News, 30 March 2002.

19. Marcus Warren, "Kabul Marvels at Accuracy of Air Strikes," *Daily Telegraph*, 23 Nov. 2001.

20. Robert Olson, "Close Air Support's New Look: Strategic Assets Go Tactical," *Armed Forces Journal International* (April 2004): 46.

21. Hunter Keeter, "Pentagon Downplays Preliminary Look at Weapons Accuracy in Afghanistan," *Defense Daily*, 10 April 2002.

22. Eric Braganca, "Joint Fires Evolution," *Military Review* (Jan.–Feb. 2004): 51.

23. Ian Sample, "Military Palmtop to Cut Collateral Damage," *New Scientist*, 9 March 2002.

24. Tom Lawhead, "First Person Singular," *Journal of Electronic Defense* (Jan. 2003): 74.

25. Hunter Keeter, "PGM Funding May Top List for Short-term Military Spending Increase," *Defense Daily*, 20 Sept. 2001, 5; and Vago Muradian, "DoD Asks Industry to Prepare for Surge; Focus on Precision Munitions," *Defense Daily*, 17 Sept. 2001.

26. The actual figure was quite classified, as JDAMs were considered extraordinarily important to the coming war effort. General Richard Myers, chairman of the Joint Chiefs of Staff, noted this in a briefing at the Pentagon on 28 March 2002 when he answered a reporter's question by rhetorically asking, "Are we going to stand up here and tell you how many JDAMs we have? No, we are not. But I just told you we are not exhausted—what term did you use? Depleted? We're not depleted." Still, adequate information existed at the time for an estimate of the number of bomb guidance kits available. This figure was determined by analyzing a wide variety of sources, including William Arkin, "The Smart Bomb That Is Shaping U.S. Iraq Strategy," http://washingtonpost.com, 18 Sept. 2002; "Boeing Awarded $378 Million Contract for Accelerated JDAM Production" (13 Sept. 2002), "Boeing Receives $235 Million JDAM Contract" (4 April 2001), and "Boeing Awarded $162 Million for JDAM Production" (25 Feb. 2000), press releases at http://www.boeing.com/defense-space/missiles/jdam; "JDAM Stocks Running Very Low, Says U.S.A.," *Jane's Defence Weekly* (10 May 1999); "Joint Direct Attack Munition," http://globalsecurity.org; James Wallace, "Aerospace Notebook: Boeing Bomb Factory Runs Lean and Smart," *Seattle Post-Intelligencer*, 8 May 2002; "Missouri-to-Kosovo Flights for B-2 Not a Concern to Wing Commander," *Inside the Air Force*, 2 July 1999, 12; Michael O'Hanlon, "We're

Ready to Fight Iraq," *Wall Street Journal*, 29 May 2002; Rip and Hasik, *Precision Revolution*, 207–8, 220, 375; Don Sharp, "JDAMs Delivered Two Months Early," *Journal of Aerospace and Defense Industry News* (21 May 1999); Richard Whittle, "Smart Weapon Popular: U.S. Commanders Embrace System Despite Errors," *Dallas Morning News*, 6 Dec. 2001; and assorted briefing slides from the USAF office of Operational Test and Evaluation (OT&E).

27. Gary Eason, "Weapons That Won the War: More Than Ever Before, the War in Iraq Was a Conflict Won by Precision Guided Munitions," BBC News Online, 16 April 2003, http://news.bbc.co.uk/2/hi/middle_east/2950403.stm.

28. Pentagon press briefing, 11 April 2003.

29. According to Jim Smith, Vice President for Precision Engagement at Raytheon (and a retired USAF brigadier), Paveway bombs accounted for about 40 percent of air-delivered ordnance dropped in the campaign. See Nick Cook, "Raytheon Looks at Iraq War Lessons," *Jane's Defence Weekly* (30 July 2003): 18. The most heavily used version was the 500-pound GBU-12.

30. Comments by Major General Leaf, in Michael Sirak, "Flexibility Key to Weapon Mix," *Jane's Defense Weekly* (18 June 2003): 45. It should be noted that B/UGM-109 Tomahawk cruise missiles were in rather shorter supply. In late March 2003, the U.S. Navy asked Raytheon, under contract to deliver 192 missiles for $260.5 million, whether the company could accelerate production to replace more rapidly the fleet's depleted stocks. By the end of March, the Navy had fired about 700 of the roughly 2,000 missiles in its prewar inventory. Some 1,500 of these were of the Block IIIC GPS/INS-guided configuration. Raytheon offered to increase the production rate at its factory in Tucson from 38 to 50 per month, assuming additional funding. The alternative was to await deliveries of the planned inventory of 2361 Block IV "Tactical Tomahawks," which were scheduled to begin arriving during fiscal year 2004. See Hunter Keeter, "$62.6 Billion Supplemental Includes $7.2 Billion For PGMs, Other Consumables," *Defense Daily*, 26 March 2003.

31. Lorenzo Cortes, "B-1 Crews Moved Quickly with JDAM Loads during Iraqi Freedom, Pilot Says," *Defense Daily*, 22 April 2003.

32. Hunter Keeter, "B-1 Drops 2,000-pound JDAMs in 'Iraqi Leadership' Strike," *Defense Daily*, 9 April 2003. Residents of the area later told journalists that Hussein had been living just across the street, but his exact whereabouts at the time have not been made public. See "Residents Report on Hussein Entourage," *Los Angeles Times*, 21 April 2003.

33. Lorenzo Cortes, "77th [Squadron] F-16s Used Variety of Precision Weapons for Multiple Roles During OIF," *Defense Daily*, 28 May 2003.

34. Harrier GR7s also fired 38 Raytheon AGM-65G2 Maverick missiles, and Tornado F3s fired 47 MBDA Air-Launched Anti-Radiation Missiles (ALARMs). "Royal Air Force Dropped More Than 400 Enhanced Paveway Bombs During OIF," *Defense Daily*, 8 July 2003.

35. Comments by the unit commander, Colonel A. T. Said, in David Blair, "'145 of my 150 men fled,' says Guard officer," *Daily Telegraph*, 17 March 2003.

36. Lorenzo Cortes, "Air Force F-117s Open Coalition Air Strikes with EGBU-27s," *Defense Daily*, 21 March 2003.

37. Lorenzo Cortes, "B-2s Performed Effectively during Iraqi Freedom, Pilot Says," *Defense Daily*, 9 April 2003.

38. Seth Stern, "Smart Bombs Move to Center Stage in U.S. Arsenal," *Christian Science Monitor*, 20 March 2003.

39. The USAF had even begun adding the JDAM to its F-15Cs. The service had traditionally resisted adding an air-to-ground mission to its F-15C squadrons because the additional training

time, it was thought, would detract from their ability to spend enough time training for air-to-air combat. The JDAM, however, is so easy to use that it upends this calculus. Still, not every F-15C can handle JDAMs on the underwing hardpoints—the older aircraft suffer from wing corrosion, and some have already had their wings replaced. See Lorenzo Cortes, "Air Force Studying Air-to-Ground Munitions On F-15C; Ready For F-15E/ JDAM Combo," *Defense Daily*, 20 May 2003; and Ron Laurenzo, "General: More JDAMs Needed in Pacific," *Defense Week Daily Update*, 13 Jan. 2004.

40. Ian Bostock, "Australian Air Force Nears JDAM Selection," *Jane's Defence Weekly* (26 Nov. 2003): 30.

41. Rowan Scarborough, "Israel Takes to Air to Cripple Hezbollah," *Washington Times*, 19 June 2006; U.S. Defense Security Cooperation Agency press release, 1 June 2004; Aluf Benn, "U.S. to Sell Israeli 5,000 Smart Bombs," *Haaretz*, 21 Sept. 2004.

42. Bill Kaczor, "Eglin AFB's Bargain Bomb Revolutionizing Warfare, Purchase of Military Equipment," Associated Press, 26 May 2002.

43. See Lisa Brem and Cynthia Ingols, *Implementing Acquisition Reform: A Case Study on Joint Direct Attack Munitions (JDAMs)*, Defense Systems Management College, May 1998.

44. Terry Little quoted in Brem and Ingols, *Implementing Acquisition Reform*, 7.

45. Brem and Ingols, *Implementing Acquisition Reform*, 8–9.

46. Cahlink, "Birth of a Bomb."

47. Gary Stoller, "JDAM Smart Bombs Prove to Be Accurate—And a Smart Buy," *USA Today*, 24 March 2003.

48. Bill Kaczor, "Eglin AFB's Bargain Bomb Revolutionizing Warfare, Purchase of Military Equipment," Associated Press, 27 May 2002.

49. Loren B. Thompson, *What Works? VIII. The Joint Direct Attack Munition: Making Acquisition Reform a Reality*, Lexington Institute, Oct. 1999.

50. Dominique Myers, "Acquisition Reform—Inside the Silver Bullet; A Comparative Analysis: JDAM Versus F-22," *Acquisition Review Quarterly* (Fall 2002): 319.

51. Cahlink, "Birth of a Bomb."

52. Giles Smith, Jeff Drezner, and Irving Lachow, *Assessing the Use of "Other Transactions" Authority for Prototype Projects*, DB-375 (Santa Monica: RAND, 2000), 26–29.

53. As Under Secretary of Defense Paul Kaminski put it, "We made JDAM a pilot acquisition reform program and we did that procurement again. We put in place all the acquisition reform initiatives and the commercial spec[ification] initiatives. We went from a RFP that had a 100-page work statement to an RFP with a two-page performance spec[ification]—not how to build this, but what we wanted it to do with no mil-specs required. The winning bid on that round was $18,000 a kit: less than one-half the original cost." Remarks at the Second Klein Symposium on the Management of Technology, Pennsylvania State University, 17 Sept. 1997.

54. Insights from Leonard Shapiro, past consultant to the office of the Assistant Secretary of the Air Force for Acquisition.

55. Comments by JDAM program director W. Michael Hatcher in "Joint Direct Attack Munitions Given Full-rate Production Go-ahead," *Defence Data*, 30 March 2001.

56. Muradian, "DoD Asks Industry to Prepare for Surge."

57. Christopher Carey, "Boeing Gears Up to Double Bomb-Kit Work," *St. Louis Post-Dispatch*, 23 July 2002.

58. Lorenzo Cortes, "Boeing Hits JDAM Production Rate of 3,000 per month; Sees Bright Future," *Defense Daily*, 17 Sept. 2003.

59. Cahlink, "Birth of a Bomb."

60. This problem in Boeing's military production operations stretches at least as far back as 1946. See John Sutton, *Technology and Market Structure: Theory and History* (Cambridge, MA: MIT Press, 1998), 82.

61. Stoller, "JDAM Smart Bombs Prove to Be Accurate—And a Smart Buy." The lesson was not lost on Boeing's past competitor for the contract. In early 2004, Lockheed Martin announced that its Loitering Attack Missile, a GPS/INS/infrared-guided artillery rocket, would be produced at a plant in Alabama requiring a staff of not more than seventy. See Chris Gaudet, "Lockheed Chooses LAM Plant," *Defense News*, 24 May 2004, 24.

62. Bill Gertz, "Swiss Delay of Military Parts Sparks 'Buy American' Push," *Washington Times*, 25 July 2003.

63. See Elizabeth G. Book, "Pentagon Grapples with Post-War Industrial Issues," *National Defense Magazine* (June 2003).

64. Insights from Leonard Shapiro, past consultant to the Office of the Assistant Secretary of the Air Force for Acquisition.

65. "AMSTE Team Demos Seekerless Concept, Looks to Sea Strike ," *C4I News*, 21 Aug. 2003, 1.

66. Robert Roy Britt, "Satellite-Guided Bomb Misses Target, Kills 4 Afghan Civilians," http://Space.com, 14 Oct. 2001.

67. Hunter Keeter, "JDAM Friendly Fire Issue under Investigation by Central Command," *Defense Daily*, 11 Dec. 2001.

68. For more on this question, see chapter 8 of Rip and Hasik, *Precision Revolution*.

69. Stephen Trimble, "Air Force to Bolster JDAM with Anti-Jam/Spoofing Capability, *Aerospace Daily*, 11 Dec. 2002.

70. Michael Sirak, "USAF Seeks to Combine Laser, GPS in Bomb," *Jane's Defence Weekly* (8 Oct. 2003): 5. Developing separate versions for the 1,000-pound and 500-pound bomb bodies was not a trivial matter. Around 1997, flight instability problems with the smaller tail kit on 1,000-pound weapons delayed the debut of the GBU-3X JDAM for about a year. Myers, "Acquisition Reform," 316.

71. Bob Algarotti, "Boeing Begins Flight Testing New JDAM on B-2," press release from the Boeing Company, 31 March 2003.

72. "B-2 Bomber's Ability to Deliver Smart Weapons Enhanced," *Space Daily*, 2 Oct. 2003; and "U.S. Air Force B-2 Bomber Drops 80 JDAMs in Historic Test," *Space Daily*, 22 Sept. 2003.

73. Paul K. Davis, *Analytic Architecture for Capabilities-Based Planning, Mission-System Analysis, and Transformation*, MR-1513-OSD (Santa Monica: RAND, 2002), 45–46.

74. "Boeing Wins Small Diameter Bomb Deal," *Jane's Defence Weekly* (29 Aug. 2003).

75. Chad Eric Watt, "Lockheed Small Diameter Bomb Makes Test Flight," *Orlando Business Journal*, 28 July 2003.

76. Robert Algarotti, "Boeing Selected for Small Diameter Bomb Contract," Boeing press release, 28 Aug. 2003.

77. Glenn J. Goodman Jr., "New, Smaller GPS-guided Bomb Hits Target," *ISR Journal*, 27 Aug. 2004. The baseline SDB would require successful GPS guidance at this range, as its capability against moderately hardened targets at a miss distance of just six meters would be negligible. The other option would involve a low-cost terminal guidance solution aided by midcourse GPS/INS navigation, as was foreseen for later versions of the JDAM. Comments by David Foster, contract analyst for Naval Air Systems Command, 29 Oct. 2004.

78. General Richard B. Myers, USAF, "Six Months After: The Imperatives of Operation Enduring Freedom," *Royal United Services Institution Journal* (April 2002): 13.

79. Michael Sirak, "U.S. Air Force Boosts Proposed JDAM Buy," *Jane's Defence Weekly* (17 April 2002): 4.

80. This is not to say that it alone was decisive: in 1991, Coalition ground forces destroyed more Iraqi weapons in four *days* than Coalition air forces did in six *weeks*. The Iraqis were neither stunned nor incapable of fighting. See Darryl G. Press, "The Myth of Air Power in the Persian Gulf War and the Future of Warfare," *International Security* 26, no. 2 (Fall 2001): 5–44. The effect on the Taliban in 2001 and the Iraqis in 2003, however, seems to have been more dramatic.

81. Lorenzo Cortes, "Coalition Forces Have Fired 15,000 Guided Munitions during Iraqi Freedom," *Defense Daily*, 11 April 2003. Weapon integration, or a lack thereof, was one of the contributing factors in the relatively low rate of precision weapon use against Yugoslavia in 1999. The USAF used its B-1B Lancer bombers in 1999, but did not have a precision weapon at the time for the type. This meant that the B-1B crews were dropping unguided weapons in large quantities, given the huge bomb capacity of the aircraft. The B-1Bs did drop these relatively accurately, since the bombardier can make use of the aircraft's own GPS/INS navigation system and Northup Grumman APQ-164 synthetic aperture radar. By the time of the Afghan campaign in 2001, the B-1Bs had the JDAM. See Frank Wolfe, "Bombers Used Flexible Targeting for Afghan Campaign," *Defense Daily*, 26 Oct. 2001.

82. Richard Sokolsky, Stuart Johnson, and F. Stephen Larrabee, *Persian Gulf Security: Improving Allied Military Contributions*, MR-1245-AF (Santa Monica: RAND, 2001), 86–88.

83. This figure excludes anti-radiation missiles, which are exclusively intended for suppressing air defense radars. Data were gathered from the Teal Group's 1999 World Missiles Briefing and assorted other sources.

84. Michael Sirak, "USA Eyes Enhanced JDAM Bomb Capabilities," *Jane's Defence Weekly* (27 Nov. 2002): 6. Of course, this would subject the weapons to the same considerable INS drift problem that the JSOW would encounter when GPS was unavailable.

85. Glenn W. Goodman, "Terminal Accuracy: Smart Munitions Knock Out Ground Targets with Fewer Weapons, Less Collateral Damage," *Armed Forces Journal International* (Oct. 2002): 67.

86. Hubert Joly, Jürgen Kluge, and Lothar Stein, "Europe's Structural Weakness: Restoring the Industry to Competitive Health Means Learning to Avoid the 'High-end Trap,'" *McKinsey Quarterly* 1 (1994): 33–38.

87. David Mizell, manager, Virtual Systems, Boeing, on the Augmented Reality project, in Jim Nash, "Wiring the Jet Set: Boeing Is Equipping Factory-floor Workers with a Modified VR Setup—And Rapidly Cutting the Time It Takes to Wire New Jetliners," *Wired* (Oct. 1997).

Chapter Five

1. Robert Clifford, chairman and CEO of Incat Group, builder of the wave-piercing catamaran HMAS *Jervis Bay*, quoted in Richard Scott, "Sealift in the Fast Lane: Could Commercial Fast Ferry and High-speed Container Ship Technology Be Key to Transforming Intra-theater and Trans-oceanic Sealift?" *Jane's Defence Weekly* (8 Oct. 2003): 22.

2. Oddly, Her Majesty's Government in Australia was one of very few worldwide to recognize the claim.

3. Michael Moran, "In the Navy, Size Does Matter: Smaller, Faster Ships on the Drawing Board," *MSNBC.com*, 12 April 2001.

4. "Incat Completes Naval Mission," *Marine Log*, 11 May 2001.

5. Daniel Gouré, "The Tyranny of Forward Presence," *Naval War College Review* (Summer 2001). See also Peter Dujardin, "Difficult Choices in Future for Navy: Should Fleet Grow? If So, By How Much?" *Daily Press*, 12 Oct. 2003.

6. The last completed major effort that was publicized outside the Navy may have been a project conducted by Captain Clark "Corky" Graham at the Naval Surface Warfare Center at Carderock, Maryland, in 1989–92. This architecture focused on a large, modular ship that went by various names, including "carrier dock multimission" and "carrier of large objects," the objects being such things as aircraft, smaller scout/fighter ships, and amphibious forces. See Ronald O'Rourke, "Transformation and the Navy's Tough Choices Ahead: What Are the Options for Policy Makers?" *Naval War College Review* (Winter 2001). For published discussions of this concept, see Anne Rumsey, "Navy Plans Look-a-Likes," *Defense Week*, 13 March 1989, 3; Robert Holzer, "Navy Floats Revolutionary Ship Design for Future Fleet," *Defense News*, 14 May 1990, 4, 52; Norman Polmar, "Carrying Large Objects," *Proceedings of the U.S. Naval Institute* (Dec. 1990): 121–22; Edward J. Walsh, "'Alternative Battle Force' Stresses Commonality, Capability," *Sea Power* (Feb. 1991): 33–35; and Michael L. Bosworth, "Fleet Versatility by Distributed Aviation," *Proceedings of the U.S. Naval Institute* (Jan. 1992): 99–102. See also the "USN's '2030' Plan for Future Fleet," *Sea Power* (April 1992): 79, 82. At one point in the early 1990s, the Advanced Research Projects Agency (ARPA) explored an alternative fleet architecture that included mobile offshore bases and small modular boats. For a discussion, see "ARPA Envisions Future Battle Fleet," *Navy News and Undersea Technology* (3 October 1994): 3–5.

7. Pentagon briefing on the Virginia Class Submarine Contract Award by Assistant Secretary of the Navy for Research, Development, and Acquisition John Young, 14 Aug. 2003. By the summer of 2004, things were only getting worse. Northrop Grumman's Newport News yard in particular was singled out by Young in a September 2004 letter as having experienced "significant cost growth" and schedule delays in just that year alone. See "Northrop's Submarine Work Draws Criticism from Navy: Letter Says Company's Performance Deteriorating on Newest Nuclear Vessels," *Bloomberg News*, 16 Sept. 2004.

8. J. A. C. Lewis and Damien Kemp, "Portugal Set to Buy German Submarines," *Jane's Defence Weekly* (1 Oct. 2003): 5. In fairness, the 209s would have air-independent propulsion as well, but not the commando-carrying capacity of the Swedish boats.

9. For more on Sterling engines in submarines, see Edward C. Whitman, "Air-Independent Propulsion Technology Creates a New Undersea Threat," *Undersea Warfare* (Feb. 2002).

10. The ships are quite operationally cost-effective: each has but thirty-eight in the complement and an accompanying ten-person logistics unit. Joris Janssen Lok, "Sweden Relaunches HMS *Södermanland* with Sterling AIP Propulsion," *Jane's International Defense Review* (Oct. 2003): 18. For a noble effort at demonstrating the continued relevance of submarines in lower-intensity conflict, see Paul Mitchell, "Submarines in Peacekeeping," *Canadian Forces Quarterly* (Spring 2000).

11. The Navy pursued an intriguing acquisition strategy with these not-quite-ships. Nearly total design authority was to be accorded to the contractor—the Navy planned to merely specify a few broad parameters, and expect the competing contractors to solve the design problem as best they could. The only government-furnished equipment in the program would be the weapons—all other subsystems would be up to the contractors' choice, as long as they were interoperable with C4ISR systems aboard other Navy vessels. Admittedly, the Arsenal Ship project never proceeded beyond a batch of $15 million detailed design contracts. Robert S. Leonard, Jeffrey

A. Drezner, and Geoffrey Sommer, *The Arsenal Ship Acquisition Process Experience: Contrasting and Common Impressions from the Contractor Teams and Joint Program Office*, MR-1030-DARPA (Santa Monica: RAND, 1999), xvii.

12. Peter J. Dombrowski and Andrew L. Ross, "Transforming the Navy: Punching a Feather Bed?" *Naval War College Review* (Summer 2003).

13. John D. Butler, "Coming of Age: The SSGN Concept" (part II of III), *Commander Naval Submarine Force Reserve Note* (Sept. 2002): 3.

14. Navy Announces Ohio Class SSGN Conversion Contract Award, Pentagon press release, 18 Dec. 2003.

15. Guy Gugliotta, "Radical Warship Takes Shape: Navy's DD(X) May Launch Generation of Greater Efficiency," *Washington Post*, 8 Feb. 2004.

16. Roberto Suro, "Navy Plans High-Powered New Destroyer: Propulsion Progress Compared to Jump from Sail to Steam," *Washington Post*, 7 Jan. 2000, A3. A few seagoing forces, such as the Norwegian Coast Guard, had already moved to providing every sailor his own stateroom on their newer vessels.

17. Indeed, the first six *Arleigh Bruke*–class destroyers so-equipped only began already carrying their own remotely controlled mini-minehunters in the summer of 2004. See Christoper P. Cavas, "U.S. Destroyer *Momsen* Packs Minehunter," *ISR Journal* (10 Sept. 2004); and Mark Hewish, "Littoral Warfare by Remote Control," *Jane's International Defense Review* (Sept. 2004).

18. Pentagon briefing on the Application of New Technologies to U.S. Navy Destroyers; Secretary of the Navy Richard Danzig, Rear Admiral Joseph Carnevale, and Admiral Michael Mullen; 6 Jan. 2000. The huge amount of electrical power that the electric drive makes available could also be harnessed by rail guns and directed energy weapons, should these become technically feasible. These, in turn, have a wide variety of compelling advantages. See Isaac Porche, Henry Willis, and Martin Ruszkowski, *Framework for Quantifying Uncertainty in Electric Ship Design*, DB-407-ONR (Santa Monica: RAND, 2004).

19. Unsurprisingly, for a Navy with conservative leanings, the ships continued to use nuclear-propulsion, though for very sound reasons related to operations, logistics, and the maintenance of the industrial base. See John F. Schank, John Birkler, Elichi Kamiya, Edward Keating, Michael Mattock, Malcolm MacKinnon, and Denis Rushworth, *CVX Propulsion System Decision: Industrial Base Implications of Nuclear and Non-Nuclear Options*, DB-272 (Santa Monica: RAND, 1998).

20. For details on what the Navy learned about planning and contract management from that experience, see John F. Schank, Mark V. Arena, Denis Rushworth, John Birkler, and James Chiesa, *Refueling and Complex Overhaul of the USS Nimitz: Lessons for the Future*, MR-1632 (Santa Monica: RAND, 2002).

21. Renae Merle, "New Navy Contract Trims Northrop's Profit: Unhappy with Contractor Bonus System, Service Renegotiates Overhaul Accord," *Washington Post*, 16 Dec. 2003. The Navy's problems with paying for work not quite accomplished go back at least to the 1860s. See Kurt Hackemer, "Building the Military-Industrial Relationship: The U.S. Navy and American Business, 1854–1883," *Naval War College Review* (Spring 1999).

22. Anne Marie Squeo, "Military Shipbuilder Draws Fire for High Costs," *Asian Wall Street Journal*, 26 Jan. 2001.

23. Radio message from the captain of USS *Ronald Reagan* (Newport News) to the Program Executive Officer for Carriers (Washington Navy Yard), 27 Oct. 2003.

24. Colin S. Gray, "The Coast Guard and Navy: It's Time for a National Fleet," *Naval War College Review* (Summer 2001).

25. Dan Morgan, "Proposed Ship Speeds into Gathering Storm," *Washington Post*, 6 July 2003, A5.

26. Ronald O'Rourke, *Naval Transformation: Background and Issues for Congress*, RS20787 (Washington, D.C.: Congressional Research Service, 2002), 6. For favorable commentary on the urgency of the requirement, see Martha Dey, John Heere, Richard Price, and Richard Silveira, *Navy Acquisitions: Improved Littoral War-Fighting Capabilities Needed*, GAO-01–493 (Washington, D.C.: General Accounting Office, 2001).

27. Briefing, Littoral Combat Ship: An Overview, Naval Sea Systems Command, Program Executive Office for Surface Strike, Oct. 2002. Hunting mines with a 2,500-ton vessel was a novel idea, but the speed of the ship would be very valuable in quickly dealing with arising mine threats over a wide area of operations. Christopher P. Cavas, "LCS Will Boost U.S. Navy's Minesweeping Ability," *Defense News*, 18 June 2004. See also the section on mine warfare capabilities in the *Naval Transformation Roadmap 2003: Assured Access and Power Projection from the Sea* (Washington, D.C.: Department of the Navy, 2003), 26–30.

28. *Littoral Combat Ship Flight o Preliminary Design Interim Requirements Document*, Department of the Navy, 10 Feb. 2003, 2.

29. "Navy Announces Contract Award for Design of Ship," Pentagon press release, 17 July 2003.

30. Editorial on the Littoral Combat Ship contract awards, "A Good Choice," *Defense News*, 31 May 2004.

31. Christopher P. Cavas, "U.S. Navy Picks Two LCS Teams GD, Lockheed to Build Radically Different Designs," *Defense News*, 31 May 2004; Pentagon briefing by Assistant Secretary of the Navy for Research, Development, and Acquisition John Young on the Littoral Combat Ship contract awards, 28 May 2004.

32. See Jonathan S. Wiarda, "The U.S. Coast Guard in Vietnam: Achieving Success in a Difficult War," *Naval War College Review* (Spring 1998).

33. Matthew Dolan, "Report Draws Lessons from *Cole* Attack," *Virginian-Pilot* (20 July 2002). Suicide boats were later used by Iraqi insurgents in a failed attempt to destroy the Khawr al Amaya oil terminal. They did manage to kill two U.S. sailors and a U.S. coastguardsman in a rigid-hull inflatable boat off the patrol craft USS *Firebolt*. See Dale Eisman, "Boat Bombings Herald New Style of Fighting in Waters Off Iraq,' *Virginian-Pilot* (30 April 2004).

34. Peter M. Swartz and E. D. McGrady, *A Deep Legacy: Smaller-Scale Contingencies and the Forces That Shape the Navy*, CRM 98–95.10 (Alexandria: Center for Naval Analysis, 1998).

35. Many senior officers of the U.S. Navy expect that the importance of this mission will increase significantly in the near future, but the Navy has thus far institutionally been less enthused about the mission. See Frank W. Lacroix and Irving N. Blickstein, *Forks in the Road for the U.S. Navy*, DB-409 (Santa Monica: RAND, 2003).

36. Maarten van de Voort and Kevin A. O'Brien, *Seacurity: Improving the Security of the Global Sea-Container Shipping System*, MR-1695-JRC (Leiden: RAND Europe, 2003).

37. Tim Ripley, "Middle East Maritime Embargo Patrols Maintain Pressure," *Jane's Navy International* (Jan.–Feb. 2003): 11.

38. L. Dean Simmons, "Small Combatants: Implications for Effectiveness and Cost of Navy Surface Forces," *IDA Research Summaries* (Winter 2003): 1–3.

39. Eric J. Labs, *Transforming the Navy's Surface Combatant Force* (Washington, D.C.: U.S. Congressional Budget Office, 2003).

40. "Icon Ends Dili days," *(Royal Australian) Navy News*, 14 May 2001.

41. Patrick Carlyon, "Top Cat's Nine Lives," *Bulletin (Australia)* 121, no. 8 (19 Feb. 2003).

42. "Mission Complete for *Jervis Bay*," *(Royal Australian) Navy News*, 16 April 2001.

43. "INCAT Provides High Speed Option to USA," *(Royal Australian) Navy News*, 12 Nov. 2001.

44. Aaron Patrick, "War Gives Incat Chief High Hopes," *Age*, 27 April 2003.

45. Shawnee McKain, "HSV 2 *Swift* Easily Lives Up to Its Name," *Scimitar* (the U.S. Fifth Fleet newsletter), 25 Sept. 2003, 3.

46. "Toward a Swift Change," *Defense News*, 5 April 2004.

47. "Army Catamaran Hauls Equipment Double-time," American Forces News Service, 17 Sept. 2003.

48. *Army Watercraft Master Plan and Theater Support Vessel Information Briefing*, U.S. Army Combined Arms Support Command, 22 May 2002, 5, accessed at http://www.globalsecurity.org/military/library/report/2002/CASCOM_AWMP_and_7SV.pdf.

49. *The Theater Support Vessel*, Association of the United States Army issue paper, May 2002, 2; and "Army's Fast Ferry Returns to CENTCOM," U.S. Army press release, 23 June 2004.

50. "High Speed Vessel: Adaptability, Modularity, and Flexibility for the Joint Force," U.S. Naval Warfare Development Center concept paper.

51. Littoral Combat Ship (LCS) Briefing to [the] International Community, Naval Sea Systems Command, Oct. 2002.

52. Royal Danish Navy presentation, Ministry of Defense, London, 25 Sept. 2003.

53. Indeed, at this point, the question of crew size is driven by the numbers needed for damage control and resisting sudden boardings by pirates and maritime guerillas. Mark Hewish, "Navies Ask: Is the Coast Clear? Navies Are Turning to Fast, Stealthy Ships to Take the Fight Close to an Enemy's Coastline," *Jane's International Defense Review* (Oct. 2003): 42.

54. The U.S. Navy has also had a great deal of success recently with simulation-based decision making in procurement. See Kendall King, "Bringing the Customer to the Ship Designer: LPD 17's Virtual Crew," *Program Manager* (July–Aug. 2000): 74–79.

55. Michael Valenti, "Stealth on the Water: The Swedish Navy's *Visby* Corvette Is Designed to Be Virtually Invisible in Pursuit of Hostile Submarines and Underwater Mines," *Mechanical Engineering* (Jan. 2001): 56–61.

56. Melanie Bright, Christopher F. Foss, Joris Janssen Lok, and Richard Scott, "Surviving Change: As Sweden Switches Its Emphasis from Traditional Platforms to Network-based Defense, Industry Must Remain Flexible," *Jane's Defence Weekly* (10 March 2004): 28–29.

57. Björn Allenström, "HMS *Visby*—The Future Is Here,' *SSPA Highlights* first half (2000): 7. SSPA is a marine engineering firm owned by the Chalmers University of Technology.

58. As one hull inspection officer at the U.S. Navy's Board of Inspection and Survey noted, aluminum loses structural strength rapidly as temperature increases, and quickly with exposure to those high temperatures over time. One Navy test suggested that a helicopter fire on an aluminum flight deck, if left unchecked by fire suppression systems, could cause the aircraft to fall through the deck in less than one minute.

59. Briefing on HMS *Skjold* for the U.S. Navy, Norwegian Defense Logistics Organization, Oct. 2002.

60. "Multihull Seasick . . . A New Phenomenon?" *Australian Defense Science* second half (2001).

61. David L. Rockwell, "Chaos in the Littorals: New Geographies Demand New Sensor Technologies," *Journal of Electronic Defense* (Oct. 2000).

62. Carlyon, "Top Cat's Nine Lives."

63. Bill Tuck, "The CFO Whose Job Makes Waves," *CFO* (1 Dec. 2000).

64. Mark Thompson and Simon Harrington, *Setting a Course for Australia's Naval Shipbuilding and Repair Industry* (Barton: Australian Strategic Policy Institute, 2002), 13; Tuck, "The CFO Whose Job Makes Waves."

65. Maura Angle, "Boat Builder Unhappy with Rate Cut," *World Today*, Australian Broadcasting Corporation, 4 April 2001.

66. Nicole Johnson, "U.S. Navy Throws Incat Lifeline," *PM*, Australian Broadcasting Corporation, 24 July 2001.

67. Aaron Patrick, "War Gives Incat Chief High Hopes," *Age*, 27 April 2003; Andrew Darby, "A Cat above the Average Woos U.S.," *Age*, 8 June 2003; Andrew Darby, "Choppy Seas Claim Ferry Success Story," *Sydney Morning Herald*, 23 March 2002; and "Incat Steams Back to the Docks as Receivership Ends," *Sydney Morning Herald*, 15 Feb. 2003.

68. Johnson, "U.S. Navy Throws Incat Lifeline."

69. Tuck, "The CFO Whose Job Makes Waves."

70. Ian Bostock, "Incat Takes SeaFrame Concept to Fast Cats," *Jane's International Defense Review* (Oct. 2003): 8.

71. Jason Ma, "Admiral: Most LCS Requirement Analysis Done after Decision to Build," *Inside the Navy*, 13 April 2003.

72. Pat Towell, "Design for Agile Combat Ships Reflects New Face of Sea War," *Congressional Quarterly Weekly*, 1 March 2003, 515.

73. Tuck, "The CFO Whose Job Makes Waves."

74. The major problems in this regard were the shipbuilding unions, which are politically powerful despite their lack of real work, and national procurement policies, which tended to steer work toward domestic shipyards. Marc Champion and David Pearson, "Europe's Big-Industry Push to Prove Tough: Shipyards Provide a Test For French, German Goal of Consolidating Companies," *Wall Street Journal*, 25 May 2004, A14.

75. Stephen Baker, *National Security Assessment of the U.S. Shipbuilding and Repair Industry* (Washington D.C.: U.S. Dept of Commerce, 2001), xiv, 7.

76. Thomas Content, "Keeping Above Water: Defense Spending Gives Marinette Marine a Chance to Win New Contracts," *Milwaukee Journal Sentinel*, 12 July 2003.

77. BeEm V. Le, "C4ISR Involvement with the Distributed Engineering Plant (DEP)," 135, in *Space and Naval Warfare Systems Center San Diego Biennial Review 2001* (San Diego, 2001), 135–140, available at http://www.spawar.navy.mil/sti/publications/pubs/td/3117/135.pdf.

78. Christopher Palmeri and Stan Crock, "Northrop's Heavy Artillery: Folding in Litton, TRW, and Newport News Shipbuilding Has Paid Off Mightily," *Business Week*, 8 March 2004. See also Christopher P. Cavas, "Northrop's $463 Million Bet: Upgrading Shipyards to Lure Worldwide Business," *Defense News*, 17 May 2004.

79. Malina Brown, "Top Pentagon Officials Explain Need for Speed In LCS Acquisition," *Inside the Navy*, 1 March 2004, 1.

80. Indeed, Assistant Secretary of the Navy for Research, Development and Acquisition John Young led a delegation to European shipyards in August 2004 specifically to study how

many of them had managed to turn out small warships at lower costs and on shorter schedules than would be expected in the United States. See Tom Kington, "U.S. Officials Tour European Yards," *Defense News*, 6 Sept. 2004.

81. Rich Smith, "General Dynamics, Lockheed Score," *Motley Fool*, 28 May 2004.

82. Mike Watson, *USCG Deepwater Project*, U.S. Coast Guard Headquarters briefing, 2 May 2001. Actually, the Deepwater program probably had a better chance of proceeding to its planned conclusion than some of the U.S. Navy's shipbuilding projects, because the USCG had been badly underfunded for years and its tasks had considerably expanded during the U.S. war against al Qaeda and its allies. By RAND's analysis, even after Deepwater, the USCG would still face a 50 percent shortfall in assets for all the missions that the federal government would like it to perform. See John Birkler, Brien Alkire, Robert Button, Gordon Lee, Raj Raman, John Schank, and Carl Stephens, *The U.S. Coast Guard's Deepwater Force Modernization Plan: Can It Be Accelerated? Will It Meet Changing Security Needs?* MG-114 (Santa Monica: RAND, 2004); and C. Tyler Jones, "Coast Guard Set to Rejuvenate Ailing Fleet: Deepwater Project to Change the Way the Coast Guard Does Business," *Program Manager* (May–June 1999): 6–7.

83. *Theater Support Vessel Procurement: Industrial Base Assessment of the Potential Economic and Dual Sourcing Impacts*, Bureau of Industry and Security, U.S. Department of Commerce, Dec. 2003.

84. It could also be said that the Australian Defence Forces were in less need of radical overhaul, in that they never were really built for Cold War purposes, but rather, for fighting wars of middling intensity and enforcing the peace in Asia and the Pacific. For a longer discussion, see Aldo Borgu (Australian Strategic Policy Institute), "Defense Transformation—Application and Relevance in Australia," presentation to the conference "Defense Transformation in the Asia-Pacific: Meeting the Challenge," Asia-Pacific Center for Security Studies, Honolulu, 30 March 2004.

85. Christopher P. Cavas, "Analysts Ponder Purpose of New Chinese Craft," *Defense News*, 21 May 2004.

86. See, for example, Sam Locklear, "LCS and DD(X): U.S. Navy Can't Get These Ships Fast Enough," *Defense News*, 30 Aug. 2004. Rear Admiral Locklear was the deputy for surface warfare on the Pentagon naval staff, and commander of the *Nimitz* battle group during the 2003 Iraq campaign. Tellingly, while he wrote that "I would have used an entire squadron of HSVs if they were available," he merely commented on "how much more combat capability I would have had if my strike group included one or more multi-mission DD(X) destroyers." These are not the same thing, in two respects. First, the HSVs represented a militarily proven design, while the DDXs were essentially concept drawings at that point. Further, as the admiral implicitly admitted, the HSVs would have been immediately useful, while the DDXs would only have been particularly valuable if a sufficiently large aviation threat emerged.

87. "HSV-2 Proving to Be Prototype for Littoral Combat Ship Program," Naval Sea Systems Command press release, 2 April 2004.

88. "High-Speed Craft Goes to Sea," *Defense News* (21 June 2004): 32.

Chapter Six

1. Major Bob Beletic of the USAF's 555th Fighter Squadron, comment in Silicon Graphics marketing materials.

2. Comments by Andrew Carrington, CEO of Cambridge Research Associates, to Michael Barr, "Software to Plan Safer Combat Flights," ABC News, 25 March 2002, available at http://abcnews.go.com/Technology/story?id=98048&page=1.

3. D. Hughs, "Advanced USAF Mission Planning System Will Serve Fighters, Bombers, and Transports," *Aviation Week and Space Technology* (10 June 1991): 52–53, 57.

4. Rip and Hasik, *Precision Revolution*, 142.

5. Interview by the author with Michael Enright of Hamilton Technology Advisors, and formerly a marketing manager with Cambridge Research Associates.

6. Product description from Cambridge Research.

7. Timothy L. Thomas, "Air Operations in Low Intensity Conflict: The Case of Chechnya," *Aerospace Power Journal* (Winter 1997).

8. Warren Bass, "The Triage of Dayton," *Foreign Affairs* (Sept.–Oct. 1998).

9. The DMA was later renamed the National Imagery and Mapping Agency (NIMA), and subsequently, the National Geospatial-Intelligence Agency (NGA).

10. Richard G. Johnson, *Negotiating the Dayton Peace Accords through Digital Maps*, U.S. Institute of Peace, Feb. 1999. Just to be difficult, the USAF referred to the Navy-developed product as the Contingency Visualization Planning Support system, or CVPS.

11. "Camber Corporation Provides Arc/Info Support to Bosnian Peace Talks," briefing, 1996.

12. Timothy L. Thomas, "Preventing Conflict through Information Technology," *Military Review* (Dec. 1998–Feb. 1999): 46.

13. Ethan Waters, "Virtual War and Peace," *Wired* (March 1996).

14. Richard Holbrooke, *To End a War* (New York: Random House, 1998), 283.

15. Johnson, *Negotiating the Dayton Peace Accords through Digital Maps*; and presentation by retired Colonel Richard Johnson, U.S. Army, to the Virtual Diplomacy conference, Washington, D.C., June 2000.

16. Waters, *Virtual War and Peace*.

17. Joseph Anselmo, "Satellite Data Plays Key Role in Bosnia Peace Treaty," *Aviation Week and Space Technology* (11 Dec. 1995).

18. John Stein Monroe, "New Mapping Tools Aid Bosnian Mission," *Federal Computer Week* (29 April 1996).

19. Pentagon briefing on Operation Allied Force by Assistant Secretary of Defense for Public Affairs Kenneth Bacon, Major General Charles Wald, USAF, and others, 7 May 1999.

20. See P. Dorian, W. N. Staynes, and M. Bolton, "The Evolution of the Flight Simulator," The Royal Aeronautical Society's Fifty Years of Simulation Conference, April 1979; Kevin Moore, "A Brief History of Aircraft Flight Simulation"(Jan. 1999), available at http://homepage.ntlworld.com/bleep/SimHist1.html; *Jane's Training and Simulation Systems* (2003).

21. John Moore, "The Smart Money Is on Intelligence: Intelligence-related Business Drives Growth for Well-positioned Systems Integrators," *Federal Computer Week*, 29 Sept. 2003.

22. Karen Walker, "Short Notice Systems for Spec Ops: U.S. Army Seeks Faster Way To Get Realistic Training," *Defense News*, 5 April 2004.

23. David McGuire, "Advanced Training Technologies Allow Pilots, Soldiers to Rehearse Missions," *Washington Post*, 17 March 2003.

24. "USAF Uses Sarnoff System to Merge Data from Two UAVs in Real Time," Sarnoff Corporation press release, 15 Sept. 2004.

25. James W. Crawley, "Navy Tries out High-tech Drills to Fine-tune Firefighting Skills," *San Diego Union Tribune*, 30 Aug. 2004.

26. Jason Sherman, "U.S. Troops Get Combat Convoy Trainer," *Defense News,* 31 Aug. 2004.

27. Michael Peck, "DARPA Sketches Futuristic 'Virtual Schoolhouse': Although the Technology Does Not Yet Exist, DARPA Believes the Concept Can Work," *National Defense* (Jan. 2004).

28. John Radke, Tom Cova, Michael F. Sheridan, Austin Troy, Mu Lan, and Russ Johnson, "Application Challenges for Geographic Information Science: Implications for Research, Education, and Policy for Emergency Preparedness and Response," *Urban and Regional Information Systems Association (URISA) Journal* (Spring 2000): 21.

29. James Kitfield, "The Simulated Revolution," *Government Executive* (Aug. 1998).

30. See Almarin Phillips, *Technology and Market Structure: A Study of the Aircraft Industry* (Lexington, MA: DC Heath, 1971).

31. Katherine V. Schinasi, Cheryl Andrew, Beverly Breen, Lily Chan, Ivy Hubler, Caron Mebane, Mike Sullivan, Sameena Nooruddin, Marie Penny Ahearn, Madhav Panwar, and Randy Zoumes, *Software Acquisitions: Stronger Management Practices Are Needed to Improve DoD's Software-Intensive Weapon Acquisitions* (Washington, D.C.: U.S. General Accounting Office, 2004).

32. M. Maier, "Architecting Principles for Systems of Systems," *Systems Engineering: Practices and Tools; Proceedings of the Sixth Annual International Symposium of the International Council on Systems Engineering* ([Seattle]: International Council on Systems Engineering, 1996), 567–574.

33. Susan E. Fisher, "Battle-tested Tech: Communications. Integrated Military Strategy Provides a Lesson for Enterprise IT," *Infoworld,* 30 May 2003. Hillen later became the U.S. assistant secretary of state for political-military affairs.

34. Comments by Jim Lewis of the Center for Strategic and International Studies, in David L. Margulis, "Military's Past Meets IT future: Generations of Military R&D Gave Life to Technology the Enterprise Takes for Granted Today," *Infoworld,* 30 May 2003; and by Richard Langley of the University of New Brunswick, in Fisher, "Battle-tested Tech."

35. Anitha Reddy, "An Eye toward the Big Leagues: AMS Officials Struggle to Move Once-Storied Contractor into Same Realm as Lockheed Martin, Northrop Grumman," *Washington Post,* 16 Feb. 2004, E01.

36. Anitha Reddy, "At AMS, Strategy Falls Short: Mockett's Time Ran Out on Big Contracts, Observers Say," *Washington Post,* 12 March 2004, E05.

37. Anitha Reddy, "As AMS CEO Exits, Reflections of What Might Have Been," *Washington Post,* 22 March 2004, E01.

38. Anitha Reddy, "Local, Canadian Firms to Buy and Split AMS: Sale Price Is $858 Million for Tech Contractor," *Washington Post,* 11 March 2004, E01.

39. See Anitha Reddy, "Defense, Security Work Pushes Up CACI Profit: Earnings for June Quarter Rose 43%," *Washington Post,* 14 Aug. 2003, E05; Renae Merle, "CACI Earnings Rise on Higher Revenue," *Washington Post,* 23 Jan. 2003; Renae Merle, "Tech, Security Sales Spur CACI, PEC Profit Growth," *Washington Post,* 23 Oct. 2002, E01; Renae Merle, "Defense Work Boosts CACI Profit by 36%," *Washington Post,* 15 Aug. 2002, E05; and Joseph C. Anselmo, "Washington Techway: A Bump in the Road," *Washington Post,* 8 May 2002.

40. Anitha Reddy, "Computer Systems Spur Growth for Contractors: List of Top Firms Is Rearranged by Acquisitions," *Washington Post,* 10 May 2004, E03; and Gopal Ratnam, "CACI Grows with Pentagon's Needs: Support Services Firm Becoming Big Player in IT Sector," *Defense News,* 3 May 2004.

41. Michael Hardy, "CACI Moves into Shadows," *Federal Computer Week,* 4 March 2003.

42. Anitha Reddy, "At CACI, Concerns about Growth: Some Worry AMS Will Drag Revenue," *Washington Post*, 15 March 2004, E01.

43. Holman W. Jenkins, Jr., "Twilight of the Software Gods," *Wall Street Journal*, 11 Feb. 2004, A19.

44. Ken Berryman and Jim Seaberg, "The Outlook for Enterprise Software," *McKinsey Quarterly* first quarter (2004).

45. Personal communication to the author by Michael Enright of Hamilton Technology Advisors, formerly a marketing manager with Cambridge Research Associates.

46. Brad Peniston, "Boeing, IBM Join to Bid on U.S. Net-Centric Work," *Defense News*, 20 Sept. 2004.

47. Kevin A. Frick and Alberto Torres, "Learning from High-Tech Deals," *McKinsey Quarterly*, first quarter (2002): 115.

48. Charles Fine, *Clockspeed: Winning Industry Control in the Age of Temporary Advantage* (Perseus, 1998).

49. Martin Libicki, "Information and Nuclear RMAs Compared," *Strategic Forum* (July 1996).

Chapter Seven

1. Tim Dyhouse, "Saving Lives, One Roadside Bomb at a Time," *VFW Magazine* (May 2006).

2. Another soldier in the same vehicle, PFC Caleb Lufkin, died three weeks later at Walter Reed Hospital of wounds sustained in the same attack. While the unclassified details are sketchy, it appears that this vehicle was penetrated by the blast. Of course, a large enough bomb will penetrate any vehicle, even a heavy tank.

3. Personal correspondence with Captain Jeff Hyde, commander of C Company, 110th Missouri Engineers, 19 July 2006.

4. James Hasik, *Professional Grade: A Working Paper on Fatalities in U.S. Army Vehicles in Iraq*, June 2006, available at http://www.jameshasik.com.

5. Interview by the author with Vernon Joynt, May 2006.

6. Michael E. Porter, *The Competitive Advantage of Nations* (New York: Free Press, 1990), 81–86.

7. Vernon Joynt, "Mine Resistant Vehicles: A Force Protection Industries Perspective" (unpublished paper, 1 July 2006).

8. Joynt, "Mine Resistant Vehicles."

9. Vernon Joynt, "MRV History and Review," paper presented at Seventh International Symposium on Technology and the Mine Problem, April 2006.

10. Vernon Joynt, "Mine Resistant Vehicles." Large land animals are a traditional source of names for armored vehicles, particularly in South Africa, Germany, and Canada. Casspir, however, was an amalgam of *Caprivi Strip* and *South African Police*, which just happened to sound like the cartoon character Casper "the friendly ghost."

11. Interview by the author with Mike Aldrich, VP of Marketing, Force Protection Industries, May 2006. The approach also led to a steadily increasing revenues for Force Protection's spare parts business. The company's Integrated Logistics Support (ILS) division took in $8.5 million in 2005, but over $40 million in 2006. In keeping with the international nature of Force Protection's team, the ILS division was run by Murray Hammick, a retired major from the British Army. See "ILS Division at Force Protection, Inc. Records Dramatic Growth in '06," *Business Wire*, 23 Nov. 2006.

12. Partly in recognition of this, the Army's concepts for its Future Combat System vehicles, which once were less than 19 tons, had exceed 24 tons as of three years into the campaign in Iraq. Jen DiMascio, "Future Combat System May Gain Weight," *Inside Defense,* 27 June 2006.

13. Interview with Joynt.

14. Scott Ervin, "Force Protection's Corporate Counsel, in 'Buffalo,'" *Army Magazine* (March 2005).

15. Hasik, *Professional Grade.*

16. Data provided by the UN Protective Force (UNPROFOR) in Bosnia.

17. Joynt, "Mine Resistant Vehicles."

18. David Axe, "Soldiers, Marines Team Up in Trailblazer Patrols," *National Defense* (April 2006); "The RSD Chubby Combats the World's Land-mine Menace," *Global Defence Review* (1997); Christine Carde, "Engineers in Afghanistan Test New Mine Detector," U.S. Defense Department press release, June 2003.

19. Greg Grant, "An Answer to IEDs," *Defense News,* 12 Sept. 2005.

20. Hasik, *Professional Grade.*

21. Interview by the author with Mike Aldrich, May 2006.

22. Marketing to the Marines brought minor product development challenges as well. The USMC's procurement officer for engineer vehicles requested that Force Protection replace the key ignition common to commercial trucks (and that had been used in the Buffalo) with a push button in the Cougars. Marines, he observed, could fight, but would lose the keys.

23. Hasik, *Professional Grade.*

24. "Cameras for Cougars in Iraq and Afghanistan," *Defense Industry Daily,* 2 June 2006.

25. Force Protection has had an interesting collection of shareholders, including Sheik Mohammed al Rashid, the emir of Dubai, and Alex Mardikian, the chief designer of Von Dutch Custom Motorcycles.

26. Corporate counsel Scott Erwin, personal communication, May 2006.

27. Interview with Mike Aldrich, May 2006.

28. Force Protection began informally referring to the cabs of its Cougars and Buffalos as "capsules" after a Defense Contract Management Agency representative used the term during a site visit. The name stuck, as it evoked the protective role of the heat shield of a space probe.

29. Hasik, *Professional Grade.*

30. Frank Vizard, "The Abrams Tank Mystery," *Scientific American* (July 2004).

31. The Fiscal Year 2007 John Warner National Defense Authorization Act moved the provisions of the Berry Amendment that dealt with domestic sourcing of specialty metals to 10 U.S.C. 2534b, and slightly expanded the list of exemptions.

32. Interview with Mike Aldrich, June 2006.

33. "Britain Buying New Land Vehicles for Iraqi and Afghan Theaters," *Defense Industry Daily,* 26 July 2006; and "Defence Secretary Orders New Vehicles for Troops in Iraq and Afghanistan," Ministry of Defence press release, 25 July 2006.

34. "Force Protection Signs Cougar Production Agreement with GDLS," *Defense Industry Daily,* 18 Nov. 2006.

35. This assessment is based on comparative tours of General Dynamics' operations at JSMC Lima in March 2005 and of BAE Systems' facility in York, Pennsylvania, in April 2005.

36. Kris Osborn, "MRAP Makers Gear Up for Orders," *Defense News,* 19 May 2007.

37. Comments recorded at the Euroskeptic blog (http://eureferendum.blogspot.com/2006/06/coffins-on-wheels.html), 26 June 2006.

38. Captain Michael Biankowski, commander of Company A, 27th Engineer Battalion, in Joe Niesen, "Armored Vehicles Stand Up to Threats Posed by Roadside Bombs," Defense Department press release, Feb. 2004.

39. Richard Pyle, "Bound for Iraq, Army Vehicles Stop Off in Manhattan," *Associated Press*, 28 June 2006; personal correspondence with Force Protection VP Mike Aldrich, 15 May 2007. In December 2006, three crewmen were killed in a bomb attack on a single Buffalo.

Chapter Eight

1. Manfred Bischoff, co-chairman, EADS, quoted in Vago Muradian and Martin Agüera, "Interview: Manfred Bischoff, EADS Co-chairman," *Defense News*, 22 March 2004.

2. Comments in Kim Burger, et al., "What Went Right?" *Jane's Defence Weekly* (30 April 2003).

3. These observations are drawn in part from Steven C. Grundman, *Implications of the War in Iraq for Defense Contractors: Toward a "Transformed" Defense-Industrial Base*, presentation to the Fifth Annual Defense and Aerospace Investor, Customer and Supplier Conference, Los Angeles, 22 Sept. 2003.

4. Zoltan J. Acs and David B. Audretsch, *Innovation and Small Firms* (Cambridge, MA: MIT Press, 1990), 103.

5. Senior manager at a large consumer products company, quoted in C. W. Kester, "Today's Options for Tomorrow's Growth," *Harvard Business Review* 62, no. 2 (March–April 1984): 157.

6. Barry R. Posen, *The Sources of Military Doctrine: France, Britain, and Germany between the World Wars* (Ithaca: Cornell University Press, 1984), 54–59.

7. Vincent Davis, *The Politics of Innovation: Patterns in Navy Cases*, Monograph Series in World Affairs, 4, no. 3 (Denver: University of Denver, 1967).

8. For an impassioned argument to the contrary, see Paul Yingling, "A Failure in Generalship," *Armed Forces Journal* (May 2007).

9. Marketing manager at Bell Helicopter, July 2006, personal communication with the author.

10. "Interview with Birgitta Böhlin, Director-General of FMV," *Military Technology* (Feb. 2000): 18.

11. Douglas S. Harned and Jerrold T. Lundquist, "What Transformation Means for the Defense Industry," *McKinsey Quarterly* third quarter (2003).

12. An exception to this rule has been Force Protection, whose marketing has been clever, well-placed, and compelling. By the standards of the industry, it has been brilliant.

13. A heightened emphasis within the U.S. Defense Department on advanced concept technology demonstration projects, particularly those proposed by contractors, as technology "on-ramps" will enable large firms to make better use of their capabilities and to become more than just order takers for the Pentagon. This idea championed by former Under Secretary of Defense for Acquisition and Technology Dr. Paul Kaminski. "Beyond Mil-Spec—The Case for Radical Reform," marketing brochure, Booz, Allen and Hamilton, Oct. 1995, 4.

14. Kevin A. Frick and Alberto Torres, "Learning from High-Tech Deals," *McKinsey Quarterly* 1 (2002): 115.

15. Standing pat can also surrender management's control over the future of the enterprise, which, shareholders' benefits aside, may not thrill the senior people who are about to lose their jobs. After all, Lord Weinstock's churlish assertion that his General Electric Company (GEC) was in the business of "manufacturing, not trading companies," guaranteed that GEC would be

the one eventually traded to British Aerospace. For GEC's long-suffering shareholders, this was probably a good outcome. See "Lord of Dullest Virtue," *Economist*, 15 April 1995, 64; and Lord Weinstock's obituary in *Economist*, 27 July 2002, 73.

16. EADS's twin chairmen swore early on to avoid this pitfall, despite the multiple cross-border opportunities for misunderstandings, or worse. See Alexander Nicoll, "A Franco-German Partnership in the Cockpit," *Financial Times*, 5 Feb. 2002, 11. The recent failings at Airbus suggest that this may have been more difficult than they had imagined.

17. Bill Sweetman, "JSF—How the Battle Was Won," *Jane's Defence Weekly* (7 Nov. 2001): 5.

18. Renae Merle, "U.S. Strips Boeing of Launches: $1 Billion Sanction over Data Stolen from Rival," *Washington Post*, 25 July 2003, A01; and Bill Mann, "When Rocket Men Go Bad," *Motley Fool*, 25 July 2003.

19. David E. Cooper, *Defense Industry Consolidation: Competitive Effects of Mergers and Acquisitions*, GAO-T/NSIAD-98–112, U.S. General Accounting Office, March 1998, 6.

20. Comments by British Defence Minister George Robertson (later NATO Secretary-General and Lord Robertson of Port Ellen) to the Confederation of British Industry, 2 Nov. 1998, reported in Michael Bell, "Leaving Portsoken: Defence Procurement in the 1980s and 1990s," *Royal United Services Institute Journal* 145, no. 4 (Aug. 2000): 30–36.

21. *Transforming the Defense Industrial Base: A Roadmap* (Washington, D.C.: Dept. of Defense, Office of the Deputy Under Secretary of Defense for Industrial Policy, 2003), available at http://www.acq.osd.mil/ip/docs/transforming_the_defense_ind_base-report_only.pdf.

22. Former Under Secretary of Defense for Acquisition, Technology, and Logistics Jacques Gansler in Vago Muradian, "Do Big U.S. Programs Stifle Innovation?" *Defense News* (10 May 2004): 10.

23. Dave Shingledecker, John Weber, and Mark Klicker, "CONOPS Development Process Recommendations," working paper of the 2003 USAF C4ISR Summit, "Transforming C4ISR into Decision Superiority," Danvers, Mass.

24. Brian Ippolito and Earll Murman, *Improving the Software Upgrade Value Stream*, Lean Aerospace Initiative working paper RP01–01 (Cambridge: MIT, 2001), available at http://lean.mit.edu/index.php?option=com_docman&task=doc_view&gid=241.

25. See International Standards Organization software engineering standard 15288.

26. Bob Brewin, "JTRS Should Promote Performance: Lack of Replacement Radios Leaves Warfighters with Old Equipment," *Federal Computer Week* (27 Sept. 2004).

27. See, for example, *Assessing the Potential for Civil-Military Integration: Technologies, Processes, and Practices*, OTA-ISS-611 (Washington, D.C: U.S. Office of Technology Assessment, 1994); *Assessing the Potential for Civil-Military Integration: Selected Case Studies*, OTA-BP-ISS-158 (Washington, D.C: U.S. Office of Technology Assessment, 1995); and Jacques Gansler, *Defense Conversion: Transforming the Arsenal of Democracy* (Boston: MIT Press, 1995).

28. Maryellen Kelley and Todd A. Watkins, "In from the Cold: Prospects for the Conversion of the Defense Industrial Base," *Science* 268 (28 April 1995): 525–32.

29. For example, a manufacturer custom-builds a new conveyor system for the United Parcel Service or a massive, high-speed paper-cutting machine for a daily newspaper. In an extensive survey, managers in the machining-intensive durable goods sector reported reaping an average of 60 percent of their revenues from their three largest customers, while some 50 percent cited six or fewer potential or actual competitors, whether or not their companies performed defense work. See Maryellen R. Kelley and Todd A. Watkins, "The Myth of the Specialized Military Contractor," *Technology Review* 98, no. 3 (April 1995): 52–58.

30. See, for example, Clive Trebilcock, "'Spin-Off' in British Economic History: Armaments and Industry, 1760–1914," *Economic History Review* 22, no. 3 (Dec. 1969): 474–490.

31. It is true that the development processes for military and commercial aircraft are quite different, particularly because combat aircraft producers have not often profitably speculated on projects. Northrop's efforts to do this in the late 1980s with the F-20 Tigershark project were a particular example of this problem. See Michael S. Mutty, *A Comparison of Military and Commercial Aircraft Development* (Industrial College of the Armed Forces, National Defense University, 1993). General Atomics, of course, stands as the stark exception to this long-held assumption.

32. Mark A. Lorell, Julia Lowell, Michael Kennedy, and Hugh P. Levaux, *Cheaper, Faster, Better?: Commercial Approaches to Weapons Acquisition*, MR-1147-AF (Santa Monica: RAND, 2000).

33. W. M. Cohen and R. C. Levin, "Empirical Studies of Innovation and Market Structure," in *Handbook of Industrial Organization*, ed. R. Schmalansee and R. Willig (Amsterdam: North Holland, 1989), 2: 1059–1107.

34. P. A. Geroski, "Innovation, Technological Opportunity, and Market Structure," *Oxford Economic Papers*, 42, no. 3 (July 1990): 586–602.

35. John Lunn, "An Empirical Analysis of Process and Product Patenting: A Simultaneous Equation Framework," *Journal of Industrial Economics* (March 1986): 319–328.

36. Robbin F. Laird, *Transformation and the Defense Industrial Base: A New Model*, Defense Horizons, 26 ([Fort McNair, Washington, D.C.: Center for Technology and National Security Policy, National Defense University, 2003).

Index